Praise for *Doing AI*

"Richard shares several powerful ideas that he has discovered by examining the intellectual history of AI from the vantage point of a business leader and advanced practitioner working on machine learning. It is a refreshing reminder that practitioners deserve to be listened to as much as, if not more than, theorists working on AI. Ultimately, *Doing AI* is an essential book for technology leaders interested in AI, business, and the intersection of the two. If you want to use AI to serve customers, get this book."

—Abe Usher, co-CEO of Black Cape

"Richard has written the book on AI that I have been threatening to write for years. Surrounded by so much hype and pseudo-philosophy, AI has become, for many, a goal in itself. However, like most solutions seeking problems, it never quite delivers the promised value. Richard challenges the prevailing thinking and has the reader refocus on what matters: solving problems. At once a thoughtful history and how-to guide, this book is a must-read for anyone interested in the truth about AI and how to actually benefit from it."

—Isaac J. Faber, PhD, Chief Data Scientist,
Artificial Intelligence (AI) Task Force

"Richard digs into the intellectual components of AI but, unlike many other books on the topic of AI, he does so from the point of view of business. *Doing AI* argues that our solutions have never been better but they are not a solution to everything—but no solution is. These technologies solve problems, often they create new problems, and sometimes distract us from our problems altogether. This is a timely book to help smart people understand what AI is and is not, what others want it to be, and ultimately what businesses need solutions to be."

—David Meerman Scott, bestselling author of
The New Rules of Marketing and PR

"When I was researching AI in grad school, we joked that it is no longer AI once the software engineer solves a real-world problem. But AI is 'sexy.' So the marketers label any semi-intelligent feature 'AI.' Rich's *Doing AI* is a down-to-earth, business-focused way of looking at what is AI that is relevant to how you do business better. Read it if you want a realistic view of how AI should be done in business."

—Peng T. Ong, Cofounder and Managing Partner at Monk's Hill Ventures

"AI is an ill-defined pursuit buoyed by misleading hype—so, someone had to drill a hole through it! This book does just that, drawing back the curtain to reveal AI's false narrative, its outlandish claim to be—or at least soon be—the ultimate one-size-fits-all solution, a silver bullet capable of solving all problems. This 'single-solution fallacy' is antithetical to prudent business. A business that buys into it is a business that fails to identify and address problems and fails to provide value to customers. Author Richard Heimann takes this on with a style so crisp, clear, and unique, it just pops off the page. He comprehensively surveys the litany of troublemakers who've misguided the world with AI mythology, but then greets this mishap with the ultimate business-savvy antidote: how to effectively identify and solve real-world problems. This book will repeatedly make you go 'hmm!' as it overhauls your thinking about AI, machine learning, and problem solving in general."

—Eric Siegel, PhD , Founder of Predictive Analytics World and instructor of "Machine Learning Leadership and Practice—End-to-End Mastery"

Doing
AI

Doing
AI

A **Business-Centric** Examination of AI Culture, Goals, and Values

RICHARD HEIMANN

Matt Holt Books
An Imprint of BenBella Books, Inc.
Dallas, TX

Matt Holt Books is an imprint of BenBella Books, Inc.
10440 N. Central Expressway, Suite 800
Dallas, TX 75231
benbellabooks.com
Send feedback to feedback@benbellabooks.com.

BenBella is a federally registered trademark.
Matt Holt and logo are trademarks of BenBella Books.

Printed in the United States of America
10 9 8 7 6 5 4 3 2 1

Library of Congress Control Number: 2021030418
ISBN 9781953295736 (trade cloth)
ISBN 9781637740071 (ebook)

Editing by Greg Brown
Copyediting by Eric Wechter
Proofreading by Denise Pangia and Greg Teague
Indexing by Amy Murphy
Text design and composition by PerfecType, Nashville, TN
Cover design by Paul McCarthy
Cover image © Shutterstock / Syifa5610
Printed by Lake Book Manufacturing

To all the problems left orphaned by solution-centric thinking.

Contents

Foreword

've known Rich for nearly half my professional life. We met when I was a newly minted Chief Technology Officer at a Fortune 500 organization, and he was an up-and-coming Data Scientist arriving at the company as a result of an acquisition.

In those days, our collaboration usually went something like this: I'd ask Rich how quickly he could get something done, and Rich would then ask me why we were doing something in the first place. Our CEO and Board wanted to know how we were going to monetize AI-based solutions (reasonable), and after all, didn't all those executives and industry analysts already define for us what problems we should solve? So why slow down and retread back into problem-definition on a constant basis?

What I didn't know at the time was how clueless I really was about how best to apply different facets of artificial intelligence (AI) to different problems. As a fast-rising entrepreneur and technology executive, my success up to that point was predicated on creating matches between engineering projects (solutions) and business needs (problems). And it was always in that order: What solution did we have, and what problem could we solve with it?

And I was exactly wrong.

Time and time again, the question of what and why would lead us in new directions. Taking principles from design thinking and agile product development, paired with deep expertise in data and AI, we could start to piece together the how. We'd uncover blind spots and biases that our customers would not only take an interest in but would reward us for uncovering. For them, operating at scale meant that papering-over a tiny flaw or accepting less than specific and measurable benefits from our work could result in millions if not billions in wasted spend, ruin careers, and potentially send companies and industry leaders spiraling toward oblivion. For those projects that had global stability implications, the stakes were even higher.

Over the past ten years, Rich has shifted my thinking, and the results have been unexpectedly profound. Today, I lead arguably the best collection of CTOs at one of the world's largest and most important AI technology companies, and the lessons that Rich taught me come into play on a daily basis. They inform me and our cross-functional teams in sales, engineering, product management, and marketing, and they help our customers through nuanced endeavors where the heat of solutions is always tempting us away from truly understanding problems.

Many of Rich's lessons underpin the successes my customers, my teams, and I have had applying AI in the real world. We have a saying in my current organization's engineering function that "the first rule of Machine Learning is that you don't need Machine Learning." It's a bit tongue in cheek, but it's a nod to the importance of problem solving. This mantra also speaks to the importance of pushing problem solvers back to the problem itself, rather than solution-centric thinking that, at best, only partially solves partial needs.

In this book, you'll get the same coaching I've received from Rich, while also reaping the benefit of the refinement and precision only possible through many years of trial and error. I can't think of a more important time to redouble our collective efforts on solving pressing, specific real-world problems. And though AI will play an important role in every industry and society at large, it's not because AI is magic. It's because innovators like you will seek out the world's most pressing problems, and start with the what and why.

—Will Grannis
Founder and Managing Director
Office of the CTO, Google

Introduction

Fall in love with the problem, not the solution.
—Uri Levine

The CEO of Alphabet, Sundar Pichai, wrote that in the next ten years humanity will shift to a world that is "AI-first."[1] Pichai says AI-first means computing becomes universally available—"be it at home, at work, in the car, or on the go—and interacting with these solutions becomes much more natural, intuitive, and, above all, *intelligent*." Pichai undoubtedly has considerable intuition into how AI can provide more information users need, when and where they need it.

In his book *Onward: How Starbucks Fought for Its Life Without Losing Its Soul*, former CEO of Starbucks Howard Schultz writes how Starbucks evolved millions of people's relationships with coffee from what they drank to where and when they drank it.[2] Similarly, the shift from what informa-

---

1. Sundar Pichai, "A Personal Google, Just for You," Google, October 4, 2016, www.blog .google/products/assistant/personal-google-just-you/.

2. Howard Schultz and Joanne Gordon, *Onward: How Starbucks Fought for Its Life Without Losing Its Soul* (New York: Rodale, 2011), 13, Kindle edition.

tion customers need to when and where they need it has been a key to Google's success, enabled by solutions like Google Pixel, Google Home, Google Translate, and Google Search.

Google has important competitive advantages, with more information to draw from to provide more information users need, when and where they need it, but many of us may be puzzled by what it means to be "AI-first."[3] In fact, it is not always clear what is meant by "AI." We interpret AI-first as though we ought to literally become solution-first without knowing why. What's more is that we conceptualize an abstract, idealized solution[4] that we place before problems and customers without fully considering whether it is wise to do so, whether the hype is true, or how solution-centricity impacts *our* business.

For example, I've seen firsthand business-minded people try to define artificial intelligence, thereby needlessly, and inversely, creating external goals for problem solving. I've seen business-minded people debate and contemplate if some arbitrary solution was "real" enough to be considered AI. However, your solution does not need external goals not granted to it by the problem it solves in order to be real. This is a type of concept creep where we expand our conceptualization of a solution to include things it does not have, or demand things of our solutions they do not need for problem solving.

As founder and managing director at the office of the CTO at Google, Will Grannis, has told me, being AI-first really means doing AI last. To me, this makes sense; because anytime we set out to do something, we always have to do something else first. After all, if being AI-first literally means solution first, then we lack problem-specific information required to even know anything about the right solution. We will also lack customer- or

---

3. Sharing a likeness to AI first, is "AI-ready" (https://www.ai.mil/blog_06_18_20 -a_roadmap_to_getting_ai_ready.html) and "AI-enabled" (https://www.c4isrnet.com /artificial-intelligence/2020/09/25/the-army-just-conducted-a-massive-test-of-its -battlefield-artificial-intelligence-in-the-desert/).

4. The terms *machine, program,* and *solution* are used interchangeably throughout this book, with a tendency toward *solution*.

market-relevant direction and will fail to align strategy with the business. To craft a relevant strategy, a business should construct those statements as problem-, customer-, or market-focused, not solution-focused. Ultimately, solution-focused AI strategies are too abstract to be useful for most of us.

After all, how do you evaluate a solution that cannot be measured positively against your problem, customer, and business? The business of a solution is not a business until after it has been evaluated positively against a problem. That is, a solution is not "real" because it has a name that includes "intelligence." A solution is only a solution after it solves a problem. Problems are only real problems when they have been evaluated positively by a customer, thereby aligning with the business. A problem is only a business problem if it has been shown to be important to a customer. Rarely is something as abstract as "intelligence" important to business, because it is rarely important to customers.[5]

Successful ventures start with a problem and a group of people who are unwilling or unable to solve that problem and therefore exist as potential customers. Consequently, businesses need to focus on solving problems and ultimately finding consumers within the context of problems, not within the context of solutions. We must reject the notion that we can find consumers within the context of a solution without a problem. Besides, every company is a technology company, and every company is a software company. Companies both large and small will have access to an arbitrary amount of data and at least some artificial intelligence or similar technology. The comprehensive understanding of problems is a competitive advantage if not the biggest (and maybe only) advantage.

To be sure, solutions have never been better, largely because of technologies like machine learning (ML). In addition to being better, solutions have never been more accessible (facilitated by open-source projects). But solutions are never just one thing; rather, they are numerous, varied,

---

5. Our collective tendency to gravitate to solutions should not come as a bombshell. We love solutions. It's a common trap for many of us. We often fall in love with someone else's solution, a sort of solution envy, or we fall in love with our own solution, a type of solution bias. However, it is hard to love solutions and still love problems for what they are.

imperfect, and certainly have not replaced problems or rendered problem solving obsolete. Although we should generally seek better solutions, our collective goal is to evaluate "better" in terms of problem solving, with problem-specific information, and progress toward a business goal, not some abstract value of "intelligence" or a vague goal of artificial intelligence.

Understanding such complicated technology in order to create the right alignment requires four things:

1) A realistic conceptualization of artificial intelligence, which is too-often hyped as *artificial minds* on an exponential path to out-thinking humans.
2) Confidence spanning boundaries that comes from identifying the goals and challenges of others, but not mistaking their goals and challenges for your own goals and challenges.
3) Goal alignment, which we must understand will not come from scholarly minded insiders who think goals originate from the study of intelligence but rather from problems and customers.
4) The ability to obtain critical business context that often cannot be found in solutions alone.

This book seeks to resolve the persistent problem of aligning so-called artificial intelligence with business by recognizing and distilling conflicting cultures, goals, and values. Focus will be on the prominent social boundary on the edge of scholarly regimes (or groups of *insiders*) as well as on practitioners (or groups of *outsiders*), where we find different groups talking about their goals through the lens of their culture and values. Specifically, we will thread a small needle and explore solutions for what they are and what they are not, what others want AI to become, why talking about an AI solution is often unlike that very solution, and, ultimately, we will explore what businesses need solutions to be.

PART ONE

UNDERSTANDING AI

SO, WHAT IS AI?
SEEING AI THROUGH
A BUSINESS LENS

By far the greatest danger of Artificial Intelligence is that
people conclude too early that they understand it.
—Eliezer Yudkowsky

S o, what is AI? And what should it mean to you, as a businessperson?
Given AI's importance and its perceived implications to harm or
do good, understanding "what it is" before we seek to "do it" seems—
on its face—important (if not necessary).

For example, in the introduction of their modern classic *Artificial
Intelligence: A Modern Approach*, authors Stuart Russell and Peter Norvig
write: "In which we try to explain why we consider artificial intelligence

to be a subject most worthy of study, and in which we try to decide what exactly it is, this being a good thing to decide before embarking." Thus, we will explore what is behind this estimable question. We will ask what exactly AI is, to whom it matters most, if there is any special significance of the name "AI" in and of itself, and how to best conceptualize AI from a business stance.

First, however, we must understand that there are those who do not want you to know exactly what AI is. David Slocum, a professor at the Berlin School of Creative Leadership, writes of consultants, cheap celebrities, and commentators who have emerged because of—and to exploit—the proliferation of information and the concurrent destabilization of standards for that information.[1] What Slocum means is that there are negative side effects associated with taking shortcuts, such as listening to the wrong people, which cost us more over time. As Yogi Berra famously said, "If you don't know where you're going, you'll end up somewhere else."

For example, consider how commentators and peevish pundits will claim an algorithm is "too dangerous to release" or "general intelligence is already here" or give a false, broad sense that so-called real AI is just around the corner.[2] Too often we depend on others to direct our attention. "Stay tuned" we are told, and we often do. We are fooled by those who willfully manipulate, and we fall into the buzzword trap, which prevents us from having the required explicit knowledge to operationalize nebulous concepts.

---

1. David Slocum, "The Innovation-Industrial Complex in the Post-Truth Era (Part 1)," *Berlin School of Creative Leadership*, January 18, 2017, www.berlin-school.com/blog/innovation-industrial-complex-post-truth-era-part-1.

2. Tom Simonite, "Open AI Said Its Code Was Risky: Two Grads Re-Created It Anyway," *Wired*, August 26, 2019, www.wired.com/story/dangerous-ai-open-source/; Aaron Krumins, "Artificial General Intelligence Is Here, and Impala Is Its Name," *ExtremeTech*, August 21, 2018, www.extremetech.com/extreme/275768-artificial-general intelligence-is-here-and-impala-is-its-name. In 2016, after the AlphaGo debut, this Slashdot quote stirred up a lot of debate: "We know now that we don't need any big new breakthroughs to get to true AI." https://games.slashdot.org/story/16/03/12/1520216/alpha-go-takes-the-match-3-0.

Ultimately, we must not exalt literary experts, who are rarely artificial intelligence experts or even practicing businesspeople, as they are safe from any responsibility. We should not glorify theorists, who may often be artificial intelligence experts but fear any responsibility of implementation and often ignore economic benefit (as well as ethical and security issues) for epistemological pursuits. We cannot extol the pseudoscientific writers who talk about what they wish solutions to be, rather than what they are. Ultimately, what you require is not compatible with these groups, their goals and values, or their lack of values. Good management starts with identifying conflicting goals and successfully spanning boundaries.

Spanning Boundaries

Business-centric thinkers and doers need to be "boundary spanners," identifying the goals and challenges of others, but not mistaking their goals and challenges for your own goals and challenges.[3] Boundary spanners are adept at bilingual translation and the ability to interpret solutions for what they are, not what someone else wants them to be. Boundary spanners solve the right problems, comprehend and communicate value propositions clearly to customers, and connect everything to the business. If you want to be an effective boundary spanner, you do not need to be both a scholar and a businessperson; you just need to understand that you are not, and often cannot be, both.

It's important to acknowledge that "AI" originated within the regime of insiders. Regimes create and support their own lexical features to describe their efforts and their goals. In other words, names like AI are useful for internal communication among insiders. As a business leader you are not a part of the insider regime but instead an outsider. You are not seeking to help AI reach its fullest potential but operationalizing how a solution, any solution, can help business success.

---

3. Michael L. Tushman, "Special Boundary Roles in the Innovation Process," *Administrative Science Quarterly* 22, no. 4 (1977): 587–605, doi:10.2307/2392402.

The hazard of adopting someone else's name occurs when the name is superficially understood, merely memorized, and casually adopted. To effectively use someone else's name, we must understand what the name represents. The Nobel Prize–winning physicist Richard Feynman articulates the difference between knowing the name of something and learning it well.[4]

"See that bird?" Feynman said. "It's a brown-throated thrush, but in Germany it's called a *halzenfugel*, and in Chinese they call it a *chung ling*, and even if you know all those names for it, you still know nothing about the bird. You only know something about people and what they call the bird."

Feynman is saying that understanding begins with a strong command of names. We must explore names and learn what they are meant to represent and to whom. As author and entrepreneur Seth Godin says, "memorizing anything that you'll need to build upon . . . is foolish," and will never yield the results you seek from them.[5]

AI pioneer Marvin Minsky calls these capacious names "suitcase words," which are words that contain a variety of meanings. In day-to-day life, suitcase words simplify communication that may prevent us from being distracted by unnecessary detail. Yet due to their usefulness and compact representations suitcase words lead to overuse. AI is a prime example. It is mainly meant to communicate complex, often amorphous, concepts to others who also understand the same complex topics. It is a shortcut for the few to talk with the few but creates friction for everyone else. The safest thing to do is simply avoid someone else's name for their work, or at the minimum, recognize what values and goals they stand for.

L. J. Rittenhouse, author of *Investing Between the Lines*, lays out her time-tested approach for recognizing at-risk businesses before trouble hits. Rittenhouse coaches corporate teams to identify blind spots and create

---

4. "The Feynman Learning Technique," *Farnam Street*, March 1, 2020, fs.blog/2015/01 /richard-feynman-knowing-something/.

5. Seth Godin, "The Difference between Memorization and Learning," *Seth's Blog*, November 9, 2019, seths.blog/2019/11/the-difference-between-memorization-and-learning/.

stakeholder communication and action strategies. The Rittenhouse Rankings have shown that CEOs with rich vocabularies gain competitive advantages by being more alert to business opportunities and business dangers. Coding executive communications allows the Rittenhouse Rankings to find business leaders "who explain complex ideas simply, balance strategic actions with vision and aspirations, and engage audiences with empathy and straight talk."[6] In other words, vocabulary mastery shapes perceptions, which in turn impacts company performance. To be sure, these high-ranking CEOs are boundary spanners.

Leaders who rank high on the Rittenhouse Rankings care about finding the right word, not just any word. Leaders who rank low don't. As Mark Twain famously observed, "The difference between the right word and the almost right word is like the difference between lightning and a lightning bug." If you use jargon to "tell it like it is," but you don't know what *it* is or the listener doesn't understand what it is, then you are actually saying that you don't know what *it* is, or you don't want someone else to know what it is. Let us avoid this circuitous journey and together find the right word, the right goals, and the right values.

The Fallacy of Definitions

First and foremost, what does everyone mean when they say, "AI"?

If there are rules to defining AI, few seem to know exactly what they are. In fact, so many definitions exist that defining AI seems to be an area of research in and of itself. The lack of universality of what AI is certainly reflects a fragmented field of study, where different uses of the same term may seriously undermine goal obtainment. This is perhaps because the name itself dares us to compare ourselves with computers and vice versa.

---

6. L. J. Rittenhouse, *Investing Between the Lines: How to Make Smarter Decisions by Decoding CEO Communications* (New York: McGraw-Hill, 2012), 43.

David Watson at the University of Oxford says that it is hard to tell whether AI is meant to be taken "literally or metaphorically."[7]

Oddly enough, the lack of agreement stems from the seemingly intuitive combination of the words *artificial* and *intelligence*. This seemingly prosaic act of taking two words that seem self-evident alone and even sound commonsensical when put together (vastly more so than economic terms like "collateralized debt obligation") proves to be double binding and sucks "adjacent ideas and images into its orbit and spaghettifies them."[8]

Part of the complication is that combining these two words is meant to invoke a new word that is deceptively powerful and allows various interpretations.

Consequently, one has to wonder how differently we would think about our technology if AI pioneers Allen Newell and Herbert Simon had won support for the significantly less hype-prone term "complex information processing," rather than "artificial intelligence," which was ultimately adopted by the field?

Newell and Simon were the only two participants at the time of the Dartmouth Workshop (considered to be the founding event for artificial intelligence as a field) who had actually developed a rudimentary "thinking machine." Perhaps they understood how tricky "intelligence" would prove to be to define and to solve. To be sure, "intelligence" has proved to become a sort of undercurrent that takes a seemingly knowable interpretation and highjacks it. How intelligence must be "solved" provides another undercurrent that exists just below the surface. Like all undercurrents, they can lead us from shore and prevent us from ever making our way back.

Let's consider a definition of artificial intelligence as "intelligence exhibited by machines." This definition can be found on Wikipedia and

---

7. David Watson, "The Rhetoric and Reality of Anthropomorphism in Artificial Intelligence," *Minds & Machines* 29 (2019): 417–40, https://doi.org/10.1007/s11023-019-09506-6.

8. John Pavlus, "Stop Pretending You Really Know What AI Is and Read This Instead," *Quartz*, September 6, 2017, qz.com/1067123/stop-pretending-you-really-know-what-ai-is-and-read-this-instead/.

exists in many similar forms.[9] Therefore, the definition is quite common, yet still pernicious. It's pernicious in the sense that this definition does not describe an area of research; rather, it attempts to describe the solution insiders seek to create. However, in that pursuit the definition is purely analytic and barely rises above tautology as the definition does not define intelligence. In fact, the definition does nothing to outline the nature of intelligence or the type of solutions we wish to call intelligent. It merely redefines intelligence as something a machine does and then points back to intelligence.

Moreover, "intelligence exhibited by machines" will never serve as a goal for insiders as it is a form of confirmation bias that is focused exclusively on positive outcomes of what a computer can do and then calling that intelligence. AI cannot be defined in such a broad way as the definition effectively makes all future research on artificial intelligence unnecessary. More aptly, AI may be defined as "machines that exhibit intelligence," where the goal is intelligence, not simply calling what a machine does intelligence. Yet, even this definition doesn't make clear what intelligence is, let alone why intelligence as a goal is desirable or even required.

A more authoritative source is an oft-cited paper by researchers Shane Legg and Marcus Hutter.[10] Legg and Hutter, both a part of Google Deep-Mind, begin their exploration by acknowledging that no general survey of definitions of intelligence have been published, which must reflect the lack of maturity in a field centrally interested in intelligence. Legg and Hutter's initial goal was merely to collate formal and informal definitions of intelligence. Compiling a definitive account would be challenging because many meanings are buried deep within articles and books. However, Legg and Hutter found seventy-odd definitions, perhaps the largest and most well referenced collection of definitions on intelligence.

---

9. More than 16,000 results on Google: https://www.google.com/search?client=safari&rls
=en&q=%E2%80%9Cintelligence+exhibited+by+machines%E2%80%9D&ie=UTF-8&oe
=UTF-8.

10. S. Legg and M. Hutter, "Universal Intelligence: A Definition of Machine Intelligence," *Minds & Machines* 17 (2007): 391–444, https://doi.org/10.1007/s11023-007-9079-x.

One of the complicating factors for defining artificial intelligence using an assortment of literary and dictionary definitions of intelligence that predate the field of artificial intelligence—as Legg and Hutter have done—is a discussion heavily weighted toward human intelligence. That is because most all definitions of intelligence that predate artificial intelligence center on a discussion of human intelligence. Thus, by using natural intelligence as a standard, other types of solutions and "intelligence" are neglected because the goal isn't intelligence in a general sense but intelligence in a specific human sense. Although there is no better kind of intelligence than human intelligence, this form of anthropocentrism creates a high, tortured standard because the goal isn't material but, rather, idealized.[11]

Consider that science fiction writer Robert Heinlein's "competent man" is sometimes seen as the archetype for insiders.[12] Heinlein notes that, "A human being should be able to change a diaper, plan an invasion, butcher a hog, conn a ship, design a building, write a sonnet, balance accounts, build a wall, set a bone, comfort the dying, take orders, give orders, cooperate, act alone, solve equations, analyze a new problem, pitch manure, program a computer, cook a tasty meal, fight efficiently, die gallantly."[13] Heinlein adds, "Specialization is for insects."

The fact that most of us can't do half of these things Heinlein considers human competence is perhaps why inspiration for artificial intelligence often comes from science fiction. This exaggerated anthropomorphic viewpoint, however, is a high standard for almost all applied problem solving. After all, businesses need to solve problems even if those solutions are as specialized as an insect's. Consider what a corresponding solution

---

11. For example, the central criticism of the Turing test is that it is explicitly anthropomorphic.

12. Will Douglas Heaven, "Artificial General Intelligence: Are We Close, and Does It Even Make Sense to Try?" *MIT Technology Review*, October 26, 2020, www.technologyreview.com/2020/10/15/1010461/artificial-general-intelligence-robots-ai-agi-deepmind-google-openai/.

13. Robert A. Heinlein, *Time Enough for Love* (New York: Ace Books, 1973), 248, Kindle edition.

would need to do. Such a solution would have to deal with incomplete and contradictory information, learn from a single example, learn in real-time, accumulate and adapt knowledge to new problems, learn autonomously, comprehend its environment by conceptualizing and understanding the real world, and do all of this with reasonable amounts of computation.[14]

Ultimately, Legg and Hutter settled on the following definition for artificial intelligence: "The goal of machine intelligence as an autonomous, goal-seeking system; [for which] intelligence measures an agent's ability to achieve goals in a wide range of environments."

Similar to other definitions, this definition explains artificial intelligence by sidestepping the whole thorny topic of intelligence; merely restating intelligence as a measure of "an agent's ability to achieve goals in a wide range of environments." It turns out that collating seventy-odd definitions of intelligence leaves us with very little information about intelligence. Or perhaps it highlights that there is nothing more to know about it?

The most practical definition of artificial intelligence was produced by one of its founders, John McCarthy: "It is the science and engineering of making intelligent machines, especially intelligent computer programs."

McCarthy's definition (like so many others) avoids explaining what intelligence is or why it is important. Instead, McCarthy's definition describes an area of research perhaps in an effort to learn more about intelligence. That said, the definition highlights an important point already made, which is that AI is not a technology. AI is an area of research meant to be an aspirational goal for someone else and, in a strict sense, not something you will build because it can't be—or at least hasn't been—built.

Undoubtedly, intelligence is a red herring of sorts for the field, because whatever it is—and insiders have not been able to define it so far—it is a multidimensional thing.[15] Intelligence is not a single thing, on a single dimen-

---

14. Peter Voss, "AGI Checklist," *Medium*, June 16, 2018, https://medium.com/intuition machine/agi-checklist-30297a4f5c1f.

15. Marvin Minsky famously described the chaotic arrangement of intelligence as the "society of the mind" in his popular book with the same name. Marvin Minsky, *The Society of Mind* (New York: Simon & Schuster, 1988).

sion, easily defined in one pithy, conclusive definition. Therefore, attempts at defining intelligence are lacking and generally adhere to the standards of the researcher and how they want a solution to be judged: not by some unified theory of intelligence or, more cynically, even what the solution claims to do. Because intelligence is too complicated to be defined, insiders include "intelligence" as a placeholder until they can learn what it is.

Although ruminating on "intelligence" is certainly a strength of the field, its vague and slippery nature makes it difficult for anyone to find much agreement on anything. Consequently, nearly all definitions of AI prove to be obscure, ambiguous, too broad, or too narrow. Ultimately, the value—especially for business—of such definitions is less than the effort that goes into acquiring them.

Definitions of artificial intelligence invariably fall into the fallacy of definitions.[16] That is, definitions fail to have merit when they are insufficiently precise. More importantly, names like "AI" fail to serve business when they are obscure, ambiguous, circular, too broad, or too narrow. Business strategy is too broad when it includes things it shouldn't, like intelligence. Business strategy is too narrow when it fails to include things that it should, like mentioning an actual problem or a real-world customer. Circular strategies are those in which a goal is defined by a solution and that solution is defined by the goal. However, you are unlikely to find a customer inside of an abstract solution. Therefore, solutions cannot be your goal, and when they are, your strategies are more complex.

The former chief data officer at Abe.ai, Francesco Gadaleta, recently explored ways AI start-ups fail.[17] Francesco's list features some classic paths to failure, including assuming AI hype is enough to succeed and operating in a technology bubble. AI, as Gadaleta writes, "cannot be built in isolation from the circumstances that make them necessary." Gadaleta adds that

---

16. "Fallacies of Definition," *Wikipedia*, July 21, 2020, en.wikipedia.org/wiki/Fallacies_of _definition.

17. F. Gadaleta, "How to Fail with Artificial Intelligence," *Medium*, April 7, 2017, https:// medium.com/money-talks-the-official-abe-blog/how-to-fail-with-artificial-intelligence -b3c4b1966bb3.

technology is not "self-sustaining," meaning you cannot prioritize technology over business.

We must conclude that there is no prima facie value in defining artificial intelligence. It is not only elusive, but also not your goal. Even when artificial intelligence is practically defined as an area of research, it reminds us that AI is not something that we can necessarily buy, or something we should even do. We should define the things that matter, like problems, value propositions, and customers. I am surprised by how frequently I am asked to help businesses define artificial intelligence, as if defining intelligence in a way that includes their solutions will make all of their work intelligent.

Not only does the name "AI" have no special significance in business, but the exercise of defining artificial intelligence is probably hazardous to business in an important way. In seeking a definition, we are explicitly adding external goals to our solutions that are not necessary for problem solving. Problem solving does not require a solution to pass external validation, and when external validation is required, problem solving is more complicated. More importantly, by creating goals for solutions, we often create an environment of advocacy for solutions, which often means we find good problems for our solutions rather than finding good problems to solve for our business.

Cognitive Plausibility

A common external goal for artificial intelligence is cognitive plausibility. That is, in order to qualify as "real," a solution must solve intelligence in much the way humans are intelligent. When it is discovered that a solution is not anthropomorphic enough, many dismiss the accomplishment. In other words, how insiders solve puzzles is as important as how they define puzzles.

In 1949, American AI pioneer Arthur Samuel designed a checkers-playing program that was among the earliest machine learning solutions.[18]

---

18. A. L. Samuel, "Some Studies in Machine Learning Using the Game of Checkers," *IBM Journal of Research and Development* 3, no. 3 (1959).

In fact, it was Samuel who coined the now well-known albeit amorphous term "machine learning," and defined it as "the ability to learn without being explicitly programmed." Samuel's checkers-playing program, which improved itself through self-play, was one of the first working instances of a machine learning solution. Samuel's program was based on the tree-search algorithm, which is still an important component for some board-playing programs today. The program was ultimately considered tricky but beatable, and insiders later determined that checkers was not a good proxy for natural intelligence, so the accomplishment was dismissed.

Chess, however, was a different story. Early pioneers of AI, including Newell and Simon, professed an exalted view of chess. In 1958, Newell and Simon wrote, "If one could devise a successful chess machine one would seem to have penetrated to the core of human intellectual endeavor."[19] As Newell and Simon highlight, chess is a skill that is linked with natural intelligence.[20] It's said that playing chess dramatically improves our ability to think rationally and develops patience and thoughtfulness. Chess is also said to help humans learn, increase cognitive skills, and further strategic thinking.

Then, in 1997, IBM's Deep Blue beat a grand master.

Deep Blue won because it found sequences of moves and searched between one hundred and two hundred million positions and possible outcomes per second.[21] Ultimately, Deep Blue showed that its performance was superior to natural intelligence by playing chess like a computer. Should we care that the solution did not play like a human, even though much of humanity, and many insiders, consider chess-playing a proxy for human intelligence? For insiders, the answer to that question is, paradoxically, a

19. A. Newell, J. C. Shaw, and H. A. Simon, "Chess Playing Programs and the Problem of Complexity," *IBM Journal of Research and Development* 2, no. 4 (1958).

20. Michigan State University, "Chess Skill Is Linked to Intelligence," *ScienceDaily*, September 13, 2016, www.sciencedaily.com/releases/2016/09/160913124722.htm.

21. Larry Greenemeier, "20 Years after Deep Blue: How AI Has Advanced Since Conquering Chess," *Scientific American*, June 2, 2017, www.scientificamerican.com/article/20 -years-after-deep-blue-how-ai-has-advanced-since-conquering-chess/.

resounding yes.[22] In other words, the goal of AI is not only to solve intelligence (whatever that is) but also to solve intelligence in a very specific way, which is the complex human way.

Humans anticipate advantages on a chessboard and work backward to find sequences of moves. This makes humans blind to alternative solutions due to limitations in cognition and the mechanization of the problem-solving set.[23] For example, a study of chess players by Professor Merim Bilalić at Northumbria University and Professor Peter McLeod at the University of Oxford that used eye-tracking technology found that even though participants believed they were looking for the fastest solution to achieve checkmate, they did not shift their gaze away from the part of the board where the familiar solution would occur.[24]

But Deep Blue plays chess like a computer, surveying the entire board at once. In fact, all computer programs play games as we'd expect computers to play games. Consider AlphaGo, which was the first computer program to defeat a professional human Go player, the first to defeat a Go world champion, and is arguably the strongest Go player in history. Yet it needed only one move to make it abundantly clear that it solved board games differently than humans.

The move happened in the second game against South Korean Lee Sedol. AlphaGo placed one of its stones on the board, which surprised everyone. "I don't really know if it's a good or bad move," said Michael Redmond, a commentator on a live English broadcast. "It's a very strange move," Redmond said. One of the Western world's best Go players, Redmond could only crack a smile. Moments later, Sedol stood from the table

---

22. This is the problem with the Wikipedia definition noted earlier. If the Wikipedia definition were correct, then Deep Blue would be considered AI, but it is not. Should we care?

23. The Einstellung Effect is the development of a mechanized state of mind often called a problem-solving set. The Einstellung Effect refers to a person's predisposition to solve problems in a specific manner even though better or more appropriate methods of solving the problem exist.

24. M. Bilalić, P. McLeod, and F. Gobet, "Inflexibility of Experts: Reality or Myth? Quantifying the Einstellung Effect in Chess Masters," *Cognitive Psychology* 56, no. 2 (March 2008): 73–102, https://doi.org/10.1016/j.cogpsych.2007.02.001.

and left the room. Sedol's next move would take nearly sixteen minutes. Visibly shaken, Sedol would lose the game and the match.

In 1957 when Herbert Simon predicted that computers would outplay humans at chess within a decade, he didn't realize he'd be off by forty years. The complexity of emerging technology may explain missing prediction targets by forty years as Simon did. However, it does not explain why after successfully beating humans at chess—actually beating Garry Kasparov, a chess grand master with the highest chess rating of *any* chess player *ever*— the feat was not considered AI by insiders.

The reason is that insiders have learned that solutions that are not cognitively plausible teach them nothing about intelligence or at least nothing more than before they started. That is, problem-specific solutions designed to *actually play to their own strengths*—strengths that are not psychologically or cognitively plausible—fall short of intelligence. This is why Deep Blue and AlphaGo are not "real" AI. The accomplishments, while profound, are achievements in problem solving, not intelligence. Nevertheless, chess-playing programs like Deep Blue have shown that the human mind can no longer claim superiority over a computer on this task. That may be chilling or exhilarating depending on your philosophical bent, but it doesn't pose an existential threat to humanity.[25]

Simon's prediction incorrectly assumed that developing a solution to solve some problem meant that the solution would require a design that looks much like how humans would otherwise solve the same problem. Simon was wrong. Simon may have also incorrectly assumed that the same solution would then develop other cognitive abilities as humans do, but this is precisely what does not happen. Instead, what happens is, like all problem solvers, insiders create shortcuts. These shortcuts are required in order to solve a problem and often takes careers to find. However, these shortcuts subvert the ostentatious goals of insiders because these efforts

---

25. After Deep Blue's victory, the 1997 *Newsweek* cover read, "The Brain's Last Stand." In the article, *Newsweek* pondered what computers would do next. Turns out they would do a lot, but Deep Blue would do nothing next. The reason is because Deep Blue is an accomplishment of problem solving, not intelligence.

produce problem-specific solutions that do not represent human cognition. Without human cognition these solutions will not solve any other problem.

Contemplate the gravity of IBM's or DeepMind's accomplishment. The human mind cannot claim superiority over a program's performance. These are certainly not perfect solutions and fall well short of real AI as they are not cognitively plausible, lack any sort of natural intelligence, and will not solve any other problems. But they are accomplishments in problem solving. The question is whether business should care that these solutions are unaware of their own accomplishment and unable to claim superiority over the human mind. The answer is no. While insiders generally want their goals to be understood and shared by their solutions, your solution does not need to understand that it is solving a problem, but you do.[26]

Solve Something or Solve Everything?

Let's revisit the definition of artificial intelligence by Legg and Hutter. We may be left wondering about whether the objective of artificial intelligence is to solve intelligence, to achieve some goal (i.e., "achieve goals"), or whether it should fulfill the mutually orthogonal goal of achieving all goals (i.e., "in a wide range of environments"). Or, all of the above?[27]

University of Oxford scholar Nick Bostrom asserts that a system's intelligence and its ability to achieve goals are orthogonal. Bostrom imagines a hypothetical artificial intelligence whose sole objective is to produce paper clips. The system's "intelligence" enables the invention of innovative ways to produce paper clips by using up all of the earth's resources. He argues

---

26. For example, DeepMind does not want to just beat humans at Go, they want their solutions to want to beat humans at Go and understand that they have accomplished the task.

27. Amazingly, I get asked whether or not the Cybraics solution can solve anything else. "AI" seems to have created a strange environment where solving a problem isn't enough. It also has to solve another problem or all problems. However, these are orthogonal pursuits. Anyone who has solved something will not solve anything else. And anyone who is trying to solve everything is solving nothing.

that "any level of intelligence could be combined with any final goal."[28] The point is subtle, but any attempt to solve "intelligence" would be orthogonal to an AI's ability to achieve a goal.[29]

Moreover, achieving all goals "in a wide range of environments" is mutually orthogonal to achieving any single goal. The reason is subtle, but when interpretation of a problem begins, insiders are no longer solving intelligence or necessarily designing solutions to acquire new skills, much less skills desirable in a wide range of environments. Instead, they are problem solving. It may sound obvious, but in order to solve a problem and accomplish some goal, interpretation of a problem is required, which will prevent you from solving all problems.[30] Consequently, problems are of little interest to insiders since they do not often, or ever, satisfy the external goals for their solutions.[31] The paradox is that insiders must decouple problems from problem solving in order to solve all problems.[32]

Let us reconsider DeepMind's AlphaGo. According to a blog post cowritten by Demis Hassabis—cofounder and CEO of DeepMind—DeepMind will no longer focus on winning Go and instead focus on "developing advanced general" solutions. Hassabis says that general solutions "could one day help scientists tackle some of our most complex problems such as

---

28. Nick Bostrom, *Superintelligence: Paths, Dangers, Strategies* (Oxford, UK: Oxford University Press, 2014), 105.

29. The tension between intelligence and goals is better known as the AI Alignment Problem. The idea is that when a specific goal is poorly constructed, or ignored over more abstract goals like solving intelligence, we must be careful with what we wish akin to *The Sorcerer's Apprentice.*

30. You may be able to find closely related problems that are solvable, but certainly not "in a wide range of environments."

31. In this way, artificial intelligence is very different from many other areas of computational research like statistics, for example, which is almost entirely applied. Statistics does not have external goals outside of application, and so problems are not a problem. You will never hear anyone ask whether or not a statistical model is "real."

32. Luciano Floridi, "A Fallacy That Will Hinder Advances in Artificial Intelligence," *Financial Times,* June 1, 2017, https://www.ft.com/content/ee996846-4626-11e7-8d27-59 b4dd6296b8.

finding new cures for diseases, dramatically reducing energy consumption, or inventing revolutionary new materials."[33]

This yearning to do more than win games is commendable. However, it also underscores that these insiders feel the need to decouple problems from problem solving in order to solve all problems. To be sure, Hassabis (a powerful insider) is less interested in solving any one of these problems (e.g., new cures for diseases, reducing energy consumption,[34] or inventing new materials) than solving all of these problems. Solving all problems rather than a single problem may sound appealing, but this type of all-or-nothing strategy is rarely effective. Most businesses cannot aspire to solve everything instead of something in particular, as most of us cannot survive on such a feast or famine strategy.[35]

DeepMind is solution solving. Solution solving is problem solving minus a problem. The reason is that DeepMind seeks a complete, general solution that may later solve real problems. In the meantime, when we ignore problems, they tend to become bigger problems.[36] Certainly, Deep-Mind's business model of solving board games as a starting point for solving all problems that may in the future solve a real problem is not a path that you need to follow. After all, ill-defined goals for solutions that "could one day" solve a problem generally have low economic value for business.

---

33. Demis Hassabis and David Silver, "AlphaGo's Next Move," DeepMind, May 27, 2017, deepmind.com/blog/article/alphagos-next-move.

34. To be fair, DeepMind has made an impact on its parent company. *See* Richard Evans and Jim Gao, "DeepMind AI Reduces Google Data Centre Cooling Bill by 40%," Deep-Mind, July 20, 2016, deepmind.com/blog/article/deepmind-ai-reduces-google-data -centre-cooling-bill-40.

35. S. Shead, "Alphabet's DeepMind Losses Soared to $570 Million in 2018," *Forbes*, August 7, 2019, https://www.forbes.com/sites/samshead/2019/08/07/deepmind-losses-soared-to -570-million-in-2018/.

36. A variant of William F. Halsey's quote, "All problems become smaller when you con-front them instead of dodging them."

The aspiration to solve everything, instead of something, is the summit for insiders,[37] and it's why they seek cognitively plausible solutions. But that does not change the fact that solutions cannot be all things to all problems, and, whether we like it or not, neither can business. Virtually no business requires solutions that are universal, because business is not universal in nature and often cannot achieve goals "in a wide range of environments."

In case there is some doubt about the possibility of a universal business, consider something simple like a go-to-market strategy (GTM). A GTM strategy focuses on how to bring new products or services to market, yet a general strategy tends to be less effective than a narrow strategy. That is, when target segments are broadly defined, prospects are easier to find, but there will be a significant variation of needs and buying behavior within each market segment. Prospects may not react consistently to the company's offering, content, or messaging mainly because customers will detect the inauthenticity of trying to be all things to all people.

The wrong prospect demands more attention and causes more trouble, often at the expense of the right customer. Ultimately, companies become optimized for their specific environment rather than all environments, and so do our solutions.[38] Additionally, a company will have access to some data and some problems but not all data or all problems. Companies should support the *ability to achieve goals*—though this does not require intelligence—but they should not consider doing so in a *wide range of environments*, because that is useless for business. What businesses require is a solution that can achieve a goal within some narrow constraints, not

---

37. Conversely, any aspiration to solve everything, instead of something, is an abyss for business.

38. The strong interpretation of this idea can be found in Clayton Christensen's so-called Innovator's Dilemma, which suggests that incumbents are less able to solve new problems than challengers since challengers would not have to destroy existing lines of business. In practice, the Innovator's Dilemma is not as restrictive as Christensen asserts, but it is certainly true that no business can concurrently solve all problems and serve every customer. Clayton M. Christensen, Stephen P. Kaufman, and Willy C. Shih, "Innovation Killers: How Financial Tools Destroy Your Capacity to Do New Things," Harvard Business Review, January 2008, https://hbr.org/2008/01/innovation-killers-how-financial-tools -destroy-your-capacity-to-do-new-things.

the design and development of a solution that can solve all problems. Said differently, don't worry if your solution cannot solve everything as long as it solves something.

The strange but important lesson for outsiders is that well-defined problems are of little interest to insiders because they do not often, or ever, satisfy the external goals for their solutions.[39] The paradox is that insiders must decouple problems from problem solving in order to solve all problems.[40] The opposite is true for business. That is, you must decouple any desire to solve intelligence from problem solving. Only when you decouple intelligence (or any external goal) from problem solving can you be successful.

If you or your team are talking about intelligence, how cognitively plausible a solution is, or if a solution can solve anything else, then you are not thinking enough about a specific problem, or the people impacted by that problem. Real-world problem solving solves the problems the world presents to intelligence whose solutions are not often—or ever—cognitively plausible. It's when you "stop trying to reproduce intelligence that you can successfully replace" or, more likely, augment it.[41]

If you have a problem to solve that aligns with a goal to achieve and seek an optimal solution to accomplish that goal, then how "cognitively plausible" some solution is, is unimportant. In fact, *how* a problem is solved is always secondary to *if* a problem is solved. The goal itself and how optimal a solution is for a problem are more important than how the goal is accomplished, much less what the solution looked like or was named after we didn't solve the problem.

Keep in mind that AI is not technology in a precise sense. It is largely a type of value alignment that does not support business-centric goal

---

39. In this way, artificial intelligence is very different from many other areas of computational research, like statistics, for example, which is almost entirely applied. Statistics does not have external goals outside of application, so problems are not a problem.

40. Floridi, "A Fallacy That Will Hinder Advances in Artificial Intelligence."

41. Floridi, "A Fallacy That Will Hinder Advances in Artificial Intelligence."

alignment. Value judgments—those judgments made to ensure value alignment—are evaluated according to whom they grant power. This power is granted to insiders, who tend to pursue perfect, cognitively plausible solutions, not problem solving. However, economic utility is for the market to decide, not peevish pundits promoting cognitively plausible solutions or extremely general learning in a broad range of environments.

The AI Effect

Marvin Minsky's definition of AI is a solution "capable of performing tasks that would require intelligence if done by humans."

Minsky's definition is more slippery than McCarthy's, which focused on AI as an area of research, because any solution that solves a problem previously performed by humans creates a new set of problems that no longer requires humans. This new, impoverished context results in new definitions of intelligence that expand to include more things done by humans until nothing humans do is considered intelligent. While this expansion lacks a humanistic approach, the issue is that as soon as a solution solves a problem, the solution is no longer artificial intelligence.

The consequence of creating external, often circular goals for solutions results in an ever-growing bull's-eye. The late computer scientist Larry Tesler called this dynamic the AI Effect. The AI Effect explains why, when insiders solve something, it ceases to be regarded as solved and their work is always what they are not doing. Similarly, solution-focused business strategies that focus on "AI" are impossible to achieve when casually adopted without the support of problems and customers.

Consider Monica Anderson, founder of Sens.AI, who says that real AI is not what is normally thought to be intelligent but rather what is normally thought *not* to be intelligent.[42] Of course, this position is quite opposite of Minsky's, who considered that AI would require intelligence, which

---

42. Monica Anderson, "Dual Process Theory," *Vimeo*, November 4, 2020, vimeo.com /111967600.

underscores the lack of internal consistency in the evaluation of goals and definitions. Nonetheless, these conflicts highlight that the discipline doesn't just have internal conflict in determining what is and what isn't intelligence but also trouble determining what aspects of intelligence are remarkable. The question we ought to ask is whether business should seek to reconcile "what is," "is not," and "what is not, is" intelligence.

Probably not.

That is, when intelligence is not intelligence, we should cease to take these discussions seriously, because we may start seeing our work as whatever we are not doing today or choose to wait to solve problems rather than using some mix of solutions insiders have declared not "real." Ultimately, outsiders need to focus on problem solving, using problem-specific information and real-world value propositions rather than solving "what is, is not, what is not, is" intelligence.

Anderson asserts that "what is not intelligent" is intuition and understanding, which ought to constitute the goal of artificial intelligence. She might be right, though it is not clear why either, much less both, are not intelligence. Nevertheless, intuition and understanding are already considered AI-hard problems that have remained elusive and unsolved for insiders.

Intuition is a function of the mind, the experience of which is described as knowledge based on a hunch, resulting (as the word itself does) from contemplation and insight. Artificial intuition would be a solution capable of making the perfect chess move, all while maintaining an awareness of its surroundings and listening for a gut-feeling to run if the room catches fire.[43]

In natural language processing (NLP), understanding would be equivalent to the reading comprehension of humans, or, in computer vision, a machine's ability to make sense of what it "sees." Although Alibaba, Microsoft, Google, and OpenAI recently developed solutions that scarcely

---

43. Professor and head of the Artificial Intelligence Lab at Stanford University, Fei-Fei Li, mentioned a variant of this line in an article from 2017. Li says the quote is from the '70s, https://www.technologyreview.com/2017/10/09/3988/put-humans-at-the-center-of-ai/. Though the quote may have first been made by Anatol Holt in the book *Organized Activity and its Support by Computer*.

(emphasis on *scarcely*) outperform humans at narrow (emphasis on *narrow*) comprehension tests, there are still many challenges. Keep in mind this isn't comprehension in the way humans think of it but rather computational comprehension. Computation produces probabilistic inference and prediction. This is everything that machines are capable of accomplishing. Yet, intelligence and computation are fundamentally different. This explains why computation will never develop into human intelligence.

Although these solutions by Alibaba, Microsoft, Google, and OpenAI regularly locate the correct answer to a question when the question and answer are co-located in a document, they struggle with longer-term dependencies where the answer is not found next to the question. Therefore, these solutions will make unreliable predictions on questions for which the correct answer is not explicitly stated. Conversely, humans can read between the lines in part because our common mode of communication is as informative as necessary and no more than is required.[44] These programs are unable to discover a meaning that is hidden or implied rather than explicitly stated, but meaning is not usually stated with the painstaking detail required to support machine understanding. These solutions are certainly interesting, economically viable even, but for real AI they are not enough.

There are many examples of dim solutions making mistakes in both understanding language and visual perception that no human would make. Consider a school bus mistaken for a snowplow or when Apple had to fix a Siri error: when told to call an ambulance, it would respond with, "Okay, from now on I'll call you 'an ambulance.'"[45]

---

44. Paul Grice, "Grice's Maxims," 1975, www.sas.upenn.edu/~haroldfs/dravling/grice .html.

45. M. A. Alcorn et al., "Strike (with) a Pose: Neural Networks Are Easily Fooled by Strange Poses of Familiar Objects," *arXiv*, November 2018, arXiv:1811.11553; Will Knight, "Tougher Turing Test Exposes Chatbots' Stupidity," *MIT Technology Review*, April 2, 2020, www.technologyreview.com/s/601897/tougher-turing-test-exposes-chatbots-stupidity/.
   Even viral solutions in machine learning like deep learning or recent advances in transformers and attention mechanisms fall short of this standard. A near miss for real AI is as good as a mile. *See* Denny Britz, "Attention and Memory in Deep Learning and NLP," *WildML*, April 27, 2016, www.wildml.com/2016/01/attention-and-memory-in-deep -learning-and-nlp/.

To be sure, AIs do not *understand* what they see or read, because they are not seeing, hearing, or reading. Our solutions certainly don't think, learn, see, read, or hear like humans. AIs do not know who "British rock group Coldplay" really is, just that it's the probabilistic answer to a question about a recent Super Bowl (for which the program also has no understanding), which makes machine learning solutions difficult and costly to build and communicating consumer value propositions tricky.

Gary Marcus writes in *Rebooting AI: Building Artificial Intelligence We Can Trust* that the real reason computers can't understand what they read or see is that they lack even a basic understanding of how the world works. For example, professor of machine learning at the University of Cambridge, Neil Lawrence, asks us to consider the six-word novel apocryphally credited to Ernest Hemingway: "For sale: baby shoes, never worn." To comprehend the significance of such a statement, humans pull together a lot of additional context. Such context is not accessible to a machine, because it does not have sufficient understanding of the human condition "to realize both the implication of the advert and what that implication means emotionally to the previous owner."[46]

Marcus's argument is akin to John Searle's famous, though much debated, "Chinese Room" thought experiment, which argues that no matter how much data a computer might be able to process, it will lack common-sense knowledge or the know-how that humans possess.[47] While this remains an open problem and illustrates how our solutions are not perfect—and in their current mode may never be perfect—it does not subvert their economic utility. It only subverts the insider goal of solving intelligence. After all, we don't need to look hard to find imperfect solutions that are useful and economically viable: consider Siri, Alexa, Google Search, Google Translate, Pandora, High-Frequency Trading, Amazon's

46. N. D. Lawrence, "The Great AI Fallacy," April 21, 2020, http://inverseprobability.com /talks/notes/the-great-ai-fallacy.html.

47. "Minds, Brains, and Programs," by John R. Searle, from *The Behavioral and Brain Sciences* 3. Copyright 1980 Cambridge University Press. Reprinted by permission of Cambridge University Press.

and Netflix's recommendations, Facebook News Feed, Flatiron Health, or Cybraics! These and many more solutions are quite useful, yet they do not truly understand anything, which may suggest that human understanding has inconsistent economic value.

The complicated problem of solving intelligence is both AI's greatest strength, because the study of intelligence is an important pursuit for humanity, and its greatest weakness, because it represents a fragmented field in which no one seems to agree on the problem or the solution. Outsiders must be aware of the AI Effect because solutions are not whatever we are not doing today but what we *are* doing today. In other words, the most important part of AI for business is the problems it solves. When all is said and done, you have to decide what success and partial success looks like, but that will never begin, or end, with intelligence synthesis.

Is AI About Simulating the Brain?[48]

The answer is sometimes, but not always. However, many insiders believe that if a solution looks like the brain, then it might actually act like the brain. When a solution doesn't act like the brain, insiders conclude that the solution teaches them nothing about the brain or intelligence.

Simulating the brain effectively requires insiders to reverse engineer it. The so-called inverse problem is the process of calculating from a set of observations the causal factors that produced them. In other words: starting with the answer and working backward to the question. However, reverse engineering the brain and studying intelligence will always be an exercise that is more complex, with much longer payoffs, than identifying and solving real-world problems.

There may be no better way of understanding ourselves than by emulating ourselves, but plagiarizing the brain by reverse engineering it is challenging, if not impossible, because the brain, neurons, and intelligence

---

48. This section is influenced by J. McCarthy, "What is AI? / Basic Questions," January 1, 1970, http://jmc.stanford.edu/artificial-intelligence/what-is-ai/index.html.

are opaque to begin with. This fact introduces noise into the description of a problem. Said differently, ill-posed, inverse problems are difficult and unstable due to ambiguous reconstruction of the answer. Even a small error can produce an extremely large error in solution reconstruction.[49]

The media and peevish pundits see the inverse proposition as some type of gift or shortcut for artificial intelligence, but the inverse proposition is rarely made to work (or at least has not yet been made to work). Hans Moravec, faculty member at Carnegie Mellon University's Robotics Institute, has argued that we should expect the difficulty of reverse engineering any human skill to be roughly proportional to the amount of time that skill has been evolving in humans. What has come to be known as Moravec's paradox highlights that what we think is easy is in fact not at all easy. The paradox is similar to the famous quote by Carl Sagan that suggests "if you wish to make an apple pie from scratch, you must first invent the universe."

Businesses have to begin with a problem and often a simple solution rather than solving a complex, ill-posed, and ill-defined one like creating artificial minds. In fact, because problem solving does not require a solution to look any specific way, a solution does not need to teach you anything about the brain, neurons,[50] or intelligence. Problem solving solves the problems the world presents to intelligence rather than studying

---

49. M. Bertero, T. A. Poggio, and V. Torre, "Ill-Posed Problems in Early Vision," *Proceedings of the IEEE* 76, no. 8 (August 1988): 869–89, doi:10.1109/5.5962.

50. The relationship between artificial neural networks and real neurons is interesting, but real neurons are much more varied than artificial neurons. Pitts & McCulloch developed the first mathematical model of neurons, which was the beginning of artificial neural networks. Warren S. McCulloch and Walter Pitts, "A Logical Calculus of the Ideas Immanent in Nervous Activity," *Bulletin of Mathematical Biophysics* 5 (1943): 115–33. Unfortunately, Pitts & McCulloch were wrong. Jerry Lettvin showed that neurons are much more complex than simple circuits or "true" or "false." "What the Frog's Eye Tells the Frog's Brain," *Proceedings of the IRE* 47, no. 11, November 1959. Although artificial neurons are a desired property of artificial neural networks, they are not easy to pin down, just as intelligence is not easy to pin down. A. L. Yuille and C. Liu, "Deep Nets: What Have They Ever Done for Vision?" *arXiv*, May 2018, arXiv:1805.04025.

intelligence. While insiders are free to study either, or both, outsiders are best served solving the problems the world presents to humans.

It would be an error for business to consider solving the ill-posed, inverse problem, though some companies like DeepMind and OpenAI are at least advocating for solving intelligence by these and other means. Hassabis described DeepMind's goal as solving intelligence and then using that to solve everything else.[51] However, business-minded leaders must understand that problems with solutions are sometimes unsolvable or at least resist solutions. American philosopher of science Thomas Kuhn emphasizes that identifying puzzles or anomalies with our solutions is a natural state for science. However, solution solving is quite unnatural for business, because trying to solve something as abstract as intelligence is not only unlikely, but will not create a customer.

Ultimately, the choice to reverse engineer the brain is a methodological one. It's right for DeepMind, OpenAI, and academic institutions because they have traditional academic habits. It's probably wrong for business. We cannot forget that their road can never be our road. Finding our road means abandoning the hope that any authority outside of us can be our master.

While simulating human cognition is an interesting methodological choice, and methods that approximate neurobiological computation like artificial neural networks have had great success, it's not a solution—or at least not the only solution—for understanding artificial intelligence. AI is incredibly broad and incredibly diverse, and businesses can pick the parts that best engender goal alignment, not the best parts that show the most promise to some subset of insiders on some disputed problem.

More importantly, solutions alone do not tell you what you need to know to determine the right problems to solve or to find the customers you need because understanding neurons, biochemistry, or someone else's brain is not enough to fully understand another human. This form

51. H. Hodson, "DeepMind and Google: The Battle to Control Artificial Intelligence," March 1, 2019, from https://www.economist.com/1843/2019/03/01/deepmind-and-google-the-battle-to-control-artificial-intelligence.

of scientism ignores that we are social animals living in a social world. As the late professor of philosophy at the University of California, Berkeley, Hubert Dreyfus pointed out, "To understand another person is not to look into the chemistry of that person's brain, not even into that person's 'soul,' but is rather to be in that person's 'shoes.'"[52] To be sure, the target domain for business is humanistic, which is to say customer focused.

Rather than asking if AI is about simulating the brain, it would be better to ask, "Are businesses required to use artificial neural networks?" If that is the question, then the answer is no. The presumption that you need to use some arbitrary solution before you identify a problem is solution guessing. Although artificial neural networks are very popular and almost perfect in the narrow sense that they can fit complex functions to data—and thus compress data into useful representations—they should never be the goal of business, because approximating a function to data is rarely enough to solve a problem and, absent of solving a problem, never the goal of business.

Finally, the reason that a solution—independent of a problem—should never be the goal of a business is that it results in mediocrity. That is, when we all use the same solutions to find problems that fit those solutions, we all tend to all find the same problems using the same solutions. Solving problems that others don't is an advantage. Using the same solution that everyone else is using to solve the same problems is not.

In fact, tomorrow's technology can quickly become today's standard. If a new technology becomes ubiquitous, and all you have is alignment to a particular solution, then you fail to differentiate yourself. Technology that seems cutting edge today will not always be so, and we must not be left holding the wrong type of alignment when technology changes.

---

52. R. Fjelland, "Why General Artificial Intelligence Will Not Be Realized," *Humanities and Social Sciences Communications* 7, no. 10 (June 2020), https://doi.org/10.1057/s41599 -020-0494-4.

2

ISN'T THERE MORE THAN ONE KIND OF AI?

The more you think you see, the less you'll actually notice.
**—Thaddeus Bradley played by Morgan
Freeman in *Now You See Me***

AI is frequently explained using the categories artificial narrow intelligence (ANI), artificial general intelligence (AGI), and artificial super-intelligence (ASI). There are few if any places where discussion of artificial intelligence occurs without these categories. Therefore, we will unpack these suspicious categories and examine this strange conceptual framework.

First and foremost, bemoaning categorizations—as I am about to do—has limited value. Categorizations are almost always wrong, or at least misleading, because they almost always include spurious information and exclude important information. Everything is arbitrarily similar and distinct, depending on how we want to classify knowledge into categories. The Ugly Duckling theorem demonstrates that swans and duckling are similar if we want to manipulate the properties for comparisons yet are also distinctly unique. In other words, all differences are equal unless we have some prior knowledge about those differences.

Artificial narrow intelligence is a subfield of what is sometimes labeled *weak* artificial intelligence. John Searle, philosopher and professor at the University of California, explained in his seminal 1980 paper, "Minds, Brains, and Programs," that weak artificial intelligence would be useful for testing hypotheses about segments of minds but would not actually be minds.[1]

Commonly accepted examples of weak- or narrow-AI include Apple's "Face ID," Amazon Alexa, Microsoft Cortana, Google Assistant, and Google Translate. These technologies are empowered by machine translation (a subfield of computational linguistics that investigates the use of software to translate text or speech from one language to another), voice recognition (an interdisciplinary subfield of computational linguistics that develops methodologies and technologies that enables the recognition and translation of spoken language into text), image classification and object detection (collectively referred to as "computer vision," which is the ability of a program to identify objects, places, people, writing, and actions in images), and NLP (a subfield of computer science, information engineering, and artificial intelligence concerned with the interactions between computers and human languages, in particular how to program computers to process and analyze large amounts of language data).

"Weak" is seen as a limited part of the mind—hence why it is often referred to as "narrow"—where any subsequent solution would ultimately

---

1. J. Searle, "Minds, Brains, and Programs," *Behavioral and Brain Sciences* 3, no. 3 (1980): 417–24. https://doi.org/10.1017/S0140525X00005756.

be limited and a superficial look-alike to natural intelligence.[2] Of course, this isn't much of a distinction from really any solution because none of our solutions have anything that closely resembles natural intelligence, and all solutions are narrow; therefore, it might be too much to call anything ANI. Since ANI is not the goal of AI, it might be too much to call these solutions "AI." Perhaps "Enterprise AI" is a better name?[3] Or no name at all? After all, naming our solutions is often nonsense.

The reason for this is because calling your work AI, or ANI, is a type of concept creep. Moreover, names like ANI will not mean for business what they mean to some insider. This is why assigning solutions someone else's name is always tricky. What matters is what problem you solve, not that a solution had a name before it solved a problem. What business requires are not better names for solutions but good, commercially viable solutions that support goal alignment. To solve sufficiently complex problems, we may use advanced technology like machine learning, but what that technology is called means less than why it is used and for whom. A business's survival doesn't rely on the name of a solution; the philosophy of AI; or a single, complete, esoteric definition of intelligence.

Artificial general intelligence (AGI) is the idealized solution many conceive when thinking about AI. AGI represents what has mostly become the single story of AI, dating back to the 1950s, with a revival in the past decade. The story, however, has fallen short of its goal, despite repeated attempts. This dogmatic pursuit to solve intelligence underscores an important point: AI is less technology and more ideology.[4]

---

2. This is similar to Marvin Minsky's definition of artificial intelligence as a solution "capable of performing tasks that would require intelligence if done by humans." That is, according to Minsky, artificial intelligence is "possible" without human intelligence; it just has to replicate something done by humans. This definition is rather broad and can be used to support everything as AI making future research obsolete.

3. It is fashionable to call business-centric "AI" Enterprise AI. That said, I am not sure it is helping any of us: https://www.google.com/search?client=safari&rls=en&q=%22 Enterprise+AI%22&ie=UTF-8&oe=UTF-8.

4. J. Lanier and G. Weyl, "AI Is an Ideology, Not a Technology," *Wired*, March, 15, 2020, https://www.wired.com/story/opinion-ai-is-an-ideology-not-a-technology/.

In 2007, Shane Legg suggested the term *AGI* to artificial intelligence researcher Ben Goertzel.[5] The same year, Goertzel and fellow researcher Cassio Pennachin popularized the name when they edited a book titled *Artificial General Intelligence*, which called for a resurgence, roughly adhering to the original vision of AI.[6]

Artificial general intelligence has since been used to represent the strong, original interpretation of AI.

Until Legg, Goertzel, and Pennachin created the name, AGI, insiders had been using "strong" and "weak" to describe their work, whereas now AGI is synonymous with strong, or real, AI.[7] Insiders ultimately working on AGI are working to solve intelligence and not some real-world problem. It could be said that no problem is too important for these insiders to solve.[8]

A few years before Goertzel edited *Artificial General Intelligence*, he was failing at solving intelligence, while also failing to solve any discernible problem. Regardless, the media coverage of Goertzel's company, Intelligenesis, and their product, Webmind, was hailed to end "mankind's reign as the sole form of reasoning and intelligence on Earth." Webmind was seen as rearing a "baby" on the internet.[9] The media declared the baby "dumb," but with a "brilliant future." Hosted on a fifty-node cluster

---

5. Should you be interested in learning more on the history of the name: https://ai .stackexchange.com/questions/20231/who-first-coined-the-term-artificial-general -intelligence.

6. Ben Goertzel and Cassio Pennachin, *Artificial General Intelligence* (New York: Springer, 2007).

7. Conceptually and philosophically speaking, AGI is not quite the same as strong-AI, nor is ANI the same as weak-AI, but they are often used interchangeably.

8. Hannah Kerner, "Too Many AI Researchers Think Real-World Problems Are Not Relevant," *MIT Technology Review*, September 9, 2020, www.technologyreview.com/2020/08 /18/1007196/ai-research-machine-learning-applications-problems-opinion/.

9. Ben Goertzel, "Waking Up from the Economy of Dreams," April 9, 2001, goertzel.org /benzine/WakingUpFromTheEconomyOfDreams.htm.

named HAL-9000,[10] after the fictional artificial intelligence character and main antagonist in Stanley Kubrick's and Arthur C. Clarke's 1968 film, *2001: A Space Odyssey*, the baby promised to change the course of human history. All the while, Intelligenesis was going bankrupt, because of their solve-no-problem business model.

The solve-no-problem business model for AI-only start-ups ignores the most valuable aspects of a business, which are identifying the problem to be solved and the value created for customers. If we look back at the last half-century, most first-generation AI companies have disappeared.[11] Peng T. Ong is managing partner at Monk's Hill Ventures, and he says it's because they weren't functional businesses to begin with. Of course, many investors were complicit. Ultimately, money was lost because solutions were evaluated by their name and how others wish solutions to operate, and because we forget that solutions create value based on what problems they solve. Most of these "businesses" avoided solving any problem.

AGI typically implies two things about a solution that should not apply to any business-minded problem solving.[12] First, that a program has the general aptitude of human intelligence (perhaps all human intelligence). Second, that an AGI like the aspirational Webmind, is a general problem solver. In other words, any knowledge of a problem is rhetorical and independent of a strategy to solve that problem. Instead, the knowledge depends on some vague, ill-defined aptitude relating to the multidimensional structure of natural intelligence, which is often meant to look like human intelligence. If that sounds grandiose, it's because it is.

Insiders have been saying for decades that AGI is just around the corner. The deeper they delve into it, however, the more they realize that it's really hard to achieve. We just need to look at how humans perceive things

---

10. This type of inspiration from science fiction should tell you all you need to know about their goals.

11. Peng T. Ong, "Why I Don't Invest in AI," *VentureBeat*, July 23, 2018, venturebeat.com /2018/07/21/why-i-dont-invest-in-ai/.

12. Stuart C. Shapiro, *Artificial Intelligence, Encyclopedia of Artificial Intelligence*, 2nd ed. (New Jersey: John Wiley), 54–57. Section 4 is on "AI-Complete Tasks."

or how we juggle between multiple unrelated thoughts and memories when making a decision to understand how complex that would be to operationalize into a program.

The truth is, only some insiders believe AGI to be a tractable problem requiring large but finite resources (especially time). Three-hundred and fifty-two insiders at the 2015 NIPS and 2015 ICML conferences were surveyed. They estimated that there's a 50 percent chance that AGI will occur by 2060. North Americans estimate a 50 percent chance of AGI in seventy-four years, on or about the year 2090. These estimates underscore some serious long-term skepticism but may still be rather bullish given our collective difficulties with confidence intervals and challenges predicting the future.[13]

Regardless, these estimates highlight that there is no AGI playbook, and trying to create one or follow some peevish pundit spells doom for all businesses. Even if no general intelligence is in sight, our solutions will not lack commercial value unless we conflate our goals with the goals of others.

A commonly held belief is that general intelligence will trigger an "intelligence explosion" that will rapidly trigger what is referred to as ASI.

Artificial super-intelligence is seen as a by-product of accomplishing the goal of AGI. It is thought that ASI is "possible" due to recursive self-improvement, the limits of which are bounded only by a program's imagination. This so-called intelligence explosion is often associated with a *technological singularity*. The singularity accelerates to meet and quickly surpass the collective intelligence of all humankind.

The only problem for ASI is that there are no more problems. When ASI solves one problem it also demands another to solve with the momentum of Newton's Cradle. An acceleration of this sort will ask itself *what is next* ad infinitum until the laws of physics or theoretical computation set in. Nick Bostrom doesn't explain how AI is supposed to work but claims that when AI becomes much smarter than the best human brains in every

---

13. Uriel Haran, "Subjective Probability Interval Estimates: A Simple and Effective Way to Reduce Overprecision in Judgment" (PhD diss., Carnegie Mellon University, 2011).

field, including scientific creativity, general wisdom, and social skills, then we've achieved ASI. A super-intelligence would also be much better at accomplishing its goals than we are at defining goals. Therefore, the real risk with ASI may not be malice but competence. As James Barrat wrote in *Our Final Invention: Artificial Intelligence and the End of the Human Era*, the debate around artificial super-intelligence may really be one around humanity's last invention.

Many believe the time between AGI and ASI will be very short. It will happen in mere months, weeks, or maybe just days—it took *Terminator*'s Skynet only twenty-five days to become "self-aware"—and the intelligence will continue to acquire knowledge at an accelerated rate.[14] This intelligence explosion was first conceived by I. J. Good,[15] the British mathematician who worked as a cryptologist at Bletchley Park with Alan Turing. Good speculated on the effects of the first ultra-intelligent machine in 1965.

"Let an ultra-intelligent machine be defined as a machine that can far surpass all the intellectual activities of any man however clever," Good said. "Since the design of machines is one of these intellectual activities, an ultra-intelligent machine could design even better machines; there would then unquestionably be an 'intelligence explosion,' and the intelligence of man would be left far behind. Thus, the first ultra-intelligent machine is the last invention that man need ever make, provided that the machine is docile enough to tell us how to keep it under control."

The intelligence explosion, however, is an unlikely outcome of artificial general intelligence, just as artificial general intelligence is an unlikely outcome of artificial narrow intelligence. In late 2017, author of *Deep Learning with Python* and Google engineer François Chollet discussed

---

14. This idea is common not only in science fiction, but it is also debated among insiders who take these notions very seriously. It is also debated with the same conviction among the media, peevish pundits, literary experts, investors, and pseudoscientific writers. In other words, this idea is not reserved for science fiction.

15. I. J. Good, "Speculations Concerning the First Ultraintelligent Machine," *Advances in Computers* 6 (1965): 31–88, https://doi.org/10.1016/S0065-2458(08)60418-0.

the implausibility of an intelligence explosion.[16] Chollet concludes that the expansion of intelligence can only come from a "co-evolution of brains (both biological and digital), sensorimotor affordances, environment, and culture not from merely tuning the gears of some brain in a jar in isolation." Chollet says that a co-evolution has already been happening and will continue as intelligence moves to an increasingly digital substrate, and he concludes that no intelligence explosion will occur, as this "process advances roughly at a linear pace."

The "brain in a jar" is the idea of a disembodied artificial intelligence. In philosophy of mind, the embodied theory of cognition holds that an agent's cognition is influenced by aspects of an agent's body beyond the brain itself.[17] Philosopher Hubert Dreyfus argued that computers could not acquire intelligence because they have no body.[18] What he meant by this was that an important part of intelligence is tacit and cannot be implemented into a computer without a body.

In their book, *The Embodied Mind*, Eleanor Rosch, Evan Thompson, and Francisco J. Varela defined *embodied* as two things: (1) cognition depends upon the kinds of experience that come from having a body with various sensorimotor capacities, and (2) these individual sensorimotor capacities are themselves embedded in a more encompassing biological, psychological, and cultural context.[19] In other words, real artificial intelli-

---

16. François Chollet, "The Implausibility of Intelligence Explosion," *Medium*, December 8, 2018, medium.com/@francois.chollet/the-impossibility-of-intelligence-explosion-5be 4a9eda6ec.

17. The extended mind thesis asks a similar question, which in its general form ponders where the mind stops, and the world begins. Andy Clark and David J. Chalmers, "The Extended Mind," *Analysis* 58, no. 1 (January 1998): 7–19. https://doi.org/10.1093/analys /58.1.7.

18. Argued in Hubert L. Dreyfus, *What Computers Still Can't Do: A Critique of Artificial Reason* (Cambridge, MA: MIT Press, 1992), 173, 235, 248, and 250, which built upon earlier work: Hubert L. Dreyfus, "Alchemy and Artificial Intelligence," Santa Monica, CA: RAND Corporation, 1965, https://www.rand.org/pubs/papers/P3244.html, and Hubert L. Dreyfus, *What Computers Can't Do* (Cambridge, MA: MIT Press, 1972).

19. Eleanor Rosch, Evan Thompson, and Francisco J. Varela, *The Embodied Mind: Cognitive Science and Human Experience* (Cambridge, MA: MIT Press, 1992), 290.

gence may only be possible if the brain in a jar has legs, which spells doom for some arbitrary GitHub repository claiming general intelligence. It also spells doom for most all businesses because aside from the esoteric nature is an ethical question that would be hard, if not impossible, to accomplish without animal experimentation and attaching legs and arms to our computers.

Although sufficiently advanced solutions like AGI and ASI present a number of concerns, not present in less-advanced systems, it may be rather unfair under the vague auspices of AI to even talk about any potential existential risk. We should not care about AI or the associated future risk. That is philosophy. There is enough risk in assuming our solutions are perfect or that they share any resemblance to human intelligence. That is precisely what they are not. Businesses need to familiarize themselves with these present-day risks and take accountability for today's solutions. Remember that the risk we have today is not from a sentient, general-intelligent machine but over-attribution of a solution's performance to human intelligence, lack of oversight and accountability, poor domain comprehension, misunderstanding the source of human error, and general misuse.[20]

Part of our collective problem when talking about AI is that we entrench our thinking in prevalent but useless dichotomies. False dichotomies are a kind of logical fallacy that creates an artificial sense that there is an alternative.

That is, ANI, AGI, and ASI suggest some sort of false balance among various technologies by presenting multiple sides of an argument that

---

20. Current examples of risk using today's technology include job loss, narrowing of labor skills, algorithmic bias, automation bias, offensive cybercampaigns, social media filter bubbles and echo chambers, targeted political ads, doctored photos and videos (i.e., deepfakes), fraud, and privacy violations. David Dao curates a list of scary usages to raise awareness. Dao outlines that even in their current state our solutions are unfair, easily susceptible to attacks, and notoriously difficult to control. David Dao: https://github.com /daviddao/awful-ai/blob/master/README.md.

don't really exist.[21] Even if we accept ANI,[22] there is no evidence to suggest anything like AGI or ASI despite the presentation of it. These categories do not delineate specific technologies, nor do they capture some kind of continuum. In fact, the inclusion of notional technology like AGI that does not exist—and may never—to evaluate today's technology uttered with a catchier name like ANI is odd. We do not often compare birds to griffins, horses to unicorns, or fish to sea serpents. Then why do we compare computation to human intelligence or to super-intelligence of all humans?

To be sure, AI is not three things. It is certainly not something that scales by "intelligence" or neatly into three bins. In fact, as a rule of thumb, the number of categories erected to explain an area of research is inversely proportional to the available knowledge about that research.

AI is one thing, and that thing, at its core, is always misaligned with business because the name implies a singular and unprecedented goal. Even if we accept that there is ANI, there is certainly no AGI or ASI, despite efforts to create false balance. As media critic Jay Rosen explains, borrowing a phrase from American philosopher Thomas Nagel, "false balance is a 'view from nowhere.'"

That said, AI is sprawling. AI is a superset that serves as an umbrella term for a constellation of other things, including abstract—almost philosophical—concepts such as understanding, intuition, reasoning, and intelligence; various areas of research such as planning, learning, natural language processing, and understanding; computer vision; and knowledge representation. AI includes macro-approaches such as machine learning (so-called sub-symbolic AI), nonstatistical methods (so-called symbolic AI), and combined learning and hybrid systems, which are a combination of different micro- and macro-approaches.

---

21. Tim Urban, "The AI Revolution: The Road to Superintelligence," Wait But Why, September 7, 2017, waitbutwhy.com/2015/01/artificial-intelligence-revolution-1.html.

22. Which is still a difference without much of a distinction.

Nonstatistical "symbolic" AI used to be the dominant paradigm of AI. Nowadays, machine learning (ML) is the dominant paradigm. But ML—like AI—is an umbrella term to generally and amorphously describe programs that learn representations without being explicitly programmed. Learning representations is the ability to learn the intrinsic structure of data by fitting a function to the data. This is a fancy way of saying pattern recognition, which is the discovery of regularities in data. *Function* is a general method by which a learning algorithm can evaluate how well a solution models the given data, which is generally done by reducing error. The error is created when the solution does not fully represent the actual relationship between variables.

ML is more technology than AI is and includes an "anarchy" of solutions.[23] For example, the ability to learn from data and fit a function is considered a critical step to support a range of tasks like classification (i.e., predicting class assignment from training data) and recommendations (i.e., predicting preferences or choices). These particular tasks are part of supervised machine learning, though ML includes various types of learning paradigms like supervised learning, reinforcement learning, unsupervised learning, semisupervised learning, and self-learning.

ML is a zoo of various micro-approaches arising from numerous families. For example, the *statistics* family of solutions includes generalized linear models such as linear regression, multiple linear regression, linear discriminant analysis, Bayesian statistics, partial least squares and principal component regression, logistic and multinomial regression, multiple adaptive regression splines, and many, many more. The *data mining* family of solutions includes rule-based classifiers, nearest-neighbors, decision trees, boosting, bagging, stacking, random forests, and other ensemble models. The *connectionists* family includes artificial neural networks and deep learning.

---

23. Joel Lehman, Jeff Clune, and Sebastian Risi, "An Anarchy of Methods: Current Trends in How Intelligence Is Abstracted in AI," *Intelligent Systems, IEEE* 29, no. 6 (November 2014): 56–62, doi:10.1109/MIS.2014.92.

In many cases, micro-approaches are specific solutions such as support vector machines, Naïve Bayes, and linear and logistic regression, while other micro-approaches—such as deep learning—aren't even specific solutions. Deep learning (DL) is yet another umbrella term for a type of architecture, specifically, artificial neural network architectures. It's also a type of methodology that explains a solution that assembles parameters with some sort of optimization method or training techniques such as gradient-based methods, for which there are dozens of variants. The abstraction of deep learning as both a general architecture and a methodology allows a variety of architectures to be applied to a variety of problems and is part of the solution's appeal and versatility. Specific and common artificial neural network architectures include convolutional neural networks, recurrent neural networks, and generative adversarial networks.

In fact, it is often a revelation for those who discover that neither artificial neural networks nor deep learning are a specific solution. Both are umbrella terms meant to describe a type of architecture, for an anarchy of methods in what may best be called an artificial neural network zoo. Fjodor van Veen, researcher at The Asimov Institute, compiled a list of more than thirty artificial neural network architectures from Attention Networks to Variational Autoencoders.[24] The artificial neural network zoo underscores something important about artificial neural networks: their true power is generated more by the number of architectures, and not the fact that they have mimicked the brain, neurons, or even how "deep" they are. What makes artificial neural networks so powerful is their flexibility in accommodating the characteristics of many problems.

Are we starting to appreciate the sprawling nature of AI?

However, any explanation that includes AGI or ASI distorts reality. Anchoring, or focalism, is a cognitive bias in which an individual relies too heavily on an initial piece of information (known as the "anchor") when

---

24. Fjodor van Veen, "Neural Network Zoo Prequel: Cells and Layers," The Asimov Institute, April 1, 2017, www.asimovinstitute.org/author/fjodorvanveen/.

making decisions. Studies have shown that anchoring is very difficult to avoid even when we are looking for it.[25] In other words, even if we recognize AGI and ASI to be significantly wrong or misplaced, they can still distort reality and create misalignments. We must not be fooled by a false dichotomy and a false balance.

25. In one study, Professors Fritz Strack and Thomas Mussweiler at the University of Würzburg gave participants anchors that were obviously wrong. They were asked whether Mahatma Gandhi died before or after age 9 or before or after age 140. Clearly neither of these options can be correct, but when the two groups were asked to suggest when they thought he had died, they guessed significantly different average ages of fifty versus sixty-seven. Fritz Strack and Thomas Mussweiler, "Explaining the Enigmatic Anchoring Effect: Mechanisms of Selective Accessibility," *Journal of Personality and Social Psychology* 73, no. 3 (1997): 437–46, https://doi.org/10.1037/0022-3514.73.3.437.

WHEN ALL YOU HAVE IS SILVER-BULLET THINKING

I suppose it is tempting, if the only tool you have is a hammer, to treat everything as if it were a nail.
—Adam Maslow

If you listen to media coverage of artificial intelligence you may believe that AI has already solved it, whatever it is. It is often implied that solutions exist before problems, which would suggest a type of silver bullet. However, media coverage about a given solution reflects only some problems (specifically problems that those solutions were meant to solve), which does not mean all problems are solved, let alone the problem that some solution was meant to solve.

When solutions exist before problems, we tend to produce strategies where a solution is our goal. However, few important problems are solved when they are found in the context of someone else's solution.[1] This type of thinking is the result of a bias called the plunging-in bias.[2] Most of us are familiar with cognitive biases ranging from anchoring to zero-sum bias,[3] but these biases are almost entirely about human decision making. The plunging-in bias is a problem-solving bias and occurs when we guess at a solution or pick problems that are good for some solution, often because we want to use some arbitrary solution so badly that we plunge into solving an arbitrary problem. This desire to finish something shapes our perception.

The fallacy is that the sooner we start solving something, the sooner we will be finished, but this is rarely the case. There are many creative ways to think about a problem and many different ways to consider a solution, depending on how problem framing occurs, but guessing and plunging is rarely one of those strategies. Impatient problem solvers run at full speed, fearing that if they spend too much time noticing or defining problems, they will miss deadlines.[4] But problems don't care when we start solving them. The goal isn't to start. It's to finish, and we can't solve a problem faster by rushing through it.

---

1. For example, since 2012, computer vision problems have been very popular problems since our awareness of computer vision solutions have been piqued by contests and benchmark data sets. More recently, solutions like GPT2/3 and BERT have made natural language processing problems fashionable. To be sure, many organizations have vision and language problems to solve, but what I have noticed is that "doing AI" is quickly conflated with these two problems, not because these problems align with business, but because they are easily found in the context of someone else's solution.

2. Gaurab Bhardwaj, Alia Crocker, Jonathan Sims, and Richard D. Wang, "Alleviating the Plunging-In Bias, Elevating Strategic Problem-Solving," *Academy of Management and Learning Education* 17, no. 3 (October 2018), https://doi.org/10.5465/amle.2017.0168.

3. Anchoring and zero-sum are meant to demonstrate the range (i.e., A–Z) of cognitive biases. That said, zero-sum bias is a cognitive bias that causes people to incorrectly perceive certain situations as zero-sum, implying that one party's gains are directly balanced by the losses of other parties.

4. Stefan Thomke and Donald Reinertsen, "Six Myths of Product Development," *Harvard Business Review*, May 2012, https://hbr.org/2012/05/six-myths-of-product-development/ar/1.

Ultimately, we must listen to a problem, and it will tell us what we need to know about the right solution. Focusing on a solution before a problem produces bad solutions with vague value propositions.[5] If you listen carefully to how successful creators discuss their daily work, they spend much more time talking about a problem even after they have found the right solution.[6] Einstein certainly didn't feel that the sooner we start the sooner we will be finished. He was known for saying—perhaps apocryphally—that if he had twenty days to solve a problem, he would take nineteen to define it. This strategy highlights that problem solving always begins with a problem, whereas problem framing continuously evolves until you discover the right solution, not the other way around where you have the right solution and continue until you discover the right problem for that solution.

In fact, you don't need to talk much about a solution if you nail problem comprehension. If you know a problem better than your target customer does (thus demonstrating a personal connection to that problem), they will assume that you know how to solve the problem. But if a customer does not believe you understand the problem, they won't give you the chance to solve it. Customers want to know if you can solve their problem, which means understanding their problem.

Often this is not what happens. Although calling something AI is great for investors and boards, it is often too solution-focused for anyone else. As an AI person, I hate solutions. No one really understands them anyway. As a businessperson, I really hate when we create goals for solutions or have a solution as a goal, without the support of a problem.

---

5. For example, Jibo, a robot from the company of the same name, took pictures and communicated with humans; however, you don't need robotics or actuators for that. All you need is a speaker and camera. Jibo was more solution than necessary because it also aspired to be a friend to humans—essentially an overpriced Alexa anthropomorphized to look more AI than it really was. Unsurprisingly, in 2019, Jibo announced that their robots would be going offline. It turns out that customers prefer a voice assistant that is a clear value proposition rather than a friend.

6. Take a look at the top fifty AI companies. They all talk about a problem more than the solution. *See* "AI 50: America's Most Promising Artificial Intelligence Companies," https://www.forbes.com/sites/jilliandonfro/2019/09/17/ai-50-americas-most-promising -artificial-intelligence-companies/?sh=7b623e84565c.

Awareness of a Solution

AI may contribute up to $15.7 trillion to the global economy by 2030.[7] If this is true, how can we explain why so few of us seem to be using AI effectively in our business?

Consider that, according to a survey of three thousand so-called AI-aware executives by McKinsey Global Institute, only one in five are using "AI" to solve problems in their business.[8] Yet, McKinsey—a leading problem-solving firm—is attempting to observe the effect of "AI" against some troubled awareness of AI.

What this survey suggests is that being AI-aware (or, more generally, solution-aware) means almost nothing if you are problem-unaware. The figure of one out of five executives underscores a point that awareness of a solution may only lead to solving some problems some of the time.

What does it mean, anyway, to be "AI-aware"? It begs the further question: Aware of what? As we know, AI is not something that is definable and thus it is not some countable occurrence.[9] In technology, there seems to be a natural, and often unquestioning inclination toward the newest thing. Our brains are attracted to novelty.[10] However, you won't know the right

---

7. According to a PWC study: https://www.pwc.com/gx/en/issues/data-and-analytics /publications/artificial-intelligence-study.html.

8. "Artificial Intelligence: The Next Digital Frontier?" (discussion paper, McKinsey & Company, June 2017), https://www.mckinsey.com/~/media/mckinsey/industries/advanced%20 electronics/our%20insights/how%20artificial%20intelligence%20can%20deliver%20 real%20value%20to%20companies/mgi-artificial-intelligence-discussion-paper.ashx.

9. Consider that artificial intelligence was mentioned 791 times during earning calls just in quarter three of 2017. How valuable is this type of "awareness" when it is disconnected from the business by being disconnected from a problem?

10. Actually, some studies have found a monotonic relationship between repetition and evaluation, where evaluation increases with repetition. Other studies have found a moderation effect where there is an inverted U-shaped relationship between exposure and affect (J. E. Crandall, V. E. Montgomery, and W. W. Reese, "Mere Exposure Versus Familiarity, with Implications for Response Competition and Expectancy Arousal Hypothesis," *Journal of General Psychology* 88 (1973): 105–20), while still others have found a novelty effect where repeated exposure leads to a decrease in affect (G. N. Cantor, "Children's 'Like-Dislike' Ratings of Familiarized and Nonfamiliarized Visual Stimuli," *Journal of Experimental Child Psychology* 6, no. 4 (1968): 651–57). Angela Y. Lee, "The Mere

solution before you understand the right problem. Moreover, we are easily misled by a solution when a problem is slightly different from someone else's (in other words, too complex, nonroutine, and unfamiliar to support our awareness of someone else's solution) to support our awareness.

Besides, what is the nature of all of this awareness? Awareness is often superficial, derived from superficial sources such as the media, journalists, literary experts, commentators, peevish pundits, and pseudoscientific writers. In the imaginative world of AI this is especially true. If awareness is influenced by dubious coverage of misinterpreted results in restricted domains, difficult or impossible to reproduce research, or vague and hyperbolic marketing statements, then our awareness exists more in some collective imagination. This can create the illusion of a perfect technology that is unlike reality and unlikely to become reality.

What results is a gap between imagination and reality. In this way, imagination subverts effort. After all, effort is necessary to achieve measurable, time-bounded goals. The larger the gap, the more frequently we will fail to create a clear vision for ourselves and our organizations. In other words, what these so-called aware executives may have is too vivid an imagination, which undercuts effort. Many choose to evaluate technology against troubled awareness, which is often dependent on some imagined solution and based on what someone else wants that solution to be rather than what it is. We need to move beyond awareness of some solution, which can be found only after solving a problem.

To be sure, imagination is what drives much of our awareness of AI because AI is largely a normative concept. In other words, AI better describes what solutions ought to be like, not what they are like. Normative thinking feels right to many when they dream of AI. The truth is few problems share universal characteristics. We cannot simply pour raw data into an AI and expect something meaningful to come out. Even if such a

Exposure Effect: Is It a Mere Case of Misattribution?" in *Advances in Consumer Research* 21, eds. Chris T. Allen and Deborah Roedder John (Provo, UT: Association for Consumer Research, 1994), 270–75.

silver bullet were to exist, thereby validating our awareness and producing an optimal solution to an arbitrary problem, it would not tell us which problems matter most, what problems to solve first, or the social, ethical, and organizational context in which problems reside. So even if we arrive at such a destination, we haven't reached the final destination.

Awareness is seductive. Talking about our perceived awareness quickly becomes as satisfying as solving problems, without any of the burden of actually—you know—solving problems. The true hazard of awareness is easy to miss. When awareness is a closed loop used to support itself, it will invariably create misalignments. Suddenly something "well said" is valued more than something "well done." The best managers understand the importance of slowing down and asking questions like, Why is change necessary? What should we change? What should it change to? and, How must we cause change? These managers have learned to pursue an awareness of problems first, not solutions. Of course, this process is superior to allowing solutions to drive business objectives. Ultimately, business needs to start with a goal in mind and understand why that goal is important.

If the numbers published by McKinsey indicate anything, it's that awareness of a solution is the very obstacle to overcome, or at least awareness of a solution is not as useful as we think or hope. One wonders if AI-unaware executives are better off than those so-called aware executives. What is the opportunity cost associated with fooling around with someone else's solutions if we ignore our problems? One of the most expensive things you can do is pay attention to the wrong thing and listen to the wrong people.

Besides, awareness and unawareness are not, as the McKinsey survey suggests, operating as some zero-sum game. That is, the acquisition of knowledge involves the revelation of ignorance; or as author Wendell Berry says, knowledge *is* the revelation of ignorance.[11] To be sure, the more we know, the more we do not know. Awareness is not carved out of our

---

11. Wendell Berry, *Standing by Words* (San Francisco: North Point Press, 1983), 72, Kindle edition.

lack of unawareness but rather highlights new areas of unawareness. Paradoxically, the most unaware of us may know more about what we don't know than the rest of us. For example, if anyone has told you they know enough about AI to be "dangerous," they are in fact dangerous. It's the classic Dunning-Kruger effect—the belief that the greatest awareness comes from the loudest voice. They'll claim to know enough about some solution to think they are right about all solutions, not realizing that they don't know enough about the right solution conditioned on the right problem.

Ultimately, awareness of a solution has very little economic value because awareness is not a benchmark for success. The truth is you will know a lot about the right solution after you need to know it. Meaning you will usually throw away more than you keep and end up learning more about a problem than your thought required.[12, 13] Business leaders are not graded on their awareness of an abstract solution or someone else's solution, but rather on how well the right problems are solved. Unfortunately, we often get carried away with awareness; where we know a lot about what others are doing but understand nothing about what problems to solve or how to solve them.

Confirming Solutions

Awareness of solutions leads to the confirmation of our awareness. This is the closed loop used to support itself that we discussed earlier in this chapter (see "Awareness of a Solution"). When a solution is the goal,

---

12. Fred Brooks wrote in *The Mythical Man-Month* to "throw the first version away," then build a second one. Brooks says that by the 20th-anniversary edition, he realized that constant incremental iteration is a far sounder approach. You build a quick prototype and get it in front of users to see what they do with it. Fred Brooks, in an interview with Kevin Kelley, *Wired*, August 2010, "The Master Planner." It's no surprise to find Fred Brooks espousing the principles of agile and lean.

13. This is why many businesses adopt lean methodologies using concepts like minimum viable products. However, even minimum viable products are solution centric using the build, measure, then learn loop. While you may be surprised what users do with your MVP, the obvious precursor to "build" is noticing and comprehending problems better.

misalignments will occur. That is, in our excitement to see how someone's solution works we implicitly align ourselves with their problem, goals, and perhaps values to mimic their work. However, confirmation of a solution is backward problem solving, where we begin with a solution and work backward to find a problem good for that solution.

Backward problem (where we first identify a solution) suggest that some problems are invisible until we are aware of a solution. But this is rarely the case. Projects and products fail more often because they solve the wrong problem with the right solution than because they get the wrong solution to the right problem. Although identifying a problem is costly and does require resources that may be difficult to acquire and allocate, it is necessary.

For example, Kentucky Fried Chicken (KFC) probably looked at advances in speech recognition and text-to-speech techniques and thought that H.A.R.L.A.N.D. (Human Assisted Robotic Linguistic Animatronic Networked Device)[14] was a good solution for these solutions. H.A.R.L.A.N.D. uses speech recognition and text-to-speech techniques to transform a KFC drive-thru operator's voice into the unmistakable drawl of Colonel Harland Sanders. It is obvious that KFC cared about the name of their solution as it shares the namesake with their founder. However, KFC didn't care enough to solve a real problem because they worked problem solving backward. When the name of a solution matters more than finding a real problem, everyone should be worried.

The solution itself, and the whole concept of an animatronic Colonel Sanders taking your fried chicken order, is more than a little goofy. The feedback on YouTube for H.A.R.L.A.N.D. was bad, and many of the comments highlight how this solution was just creepy. The solution developers ultimately leaned into the negative feedback with their own messaging, including a *Funny or Die* video that's more mockumentary than anything. H.A.R.L.A.N.D. is a lesson on the fastest way of creating a fake problem to

---

14. "KFC | H.A.R.L.A.N.D. | National Fried Chicken Day," *YouTube*, July 6, 2017, https://www.youtube.com/watch?v=shGuyx7Hu4Y&feature=youtu.be.

confirm our awareness of someone else's solution.[15] To be sure, there are enough real problems that we do not need to create new ones.[16]

If you cannot tell whether a problem is real, ask yourself if it's old. If the problem was newly created for the sole purpose of confirming a solution, then someone is likely working problem solving backward. Creating new problems because they are good for some solution is a red flag that you are no longer problem solving. New problems are sometimes fads, and they're fragile because they have not been tested by the rigor of time and may not have been shown to create or retain a customer. Finding a problem that has always been relevant is much more robust than creating a problem that was never relevant. "New" is a ripe condition for the creation of fake problems.

Instead of confirming a solution, organizations should think broadly about problems and possible solutions. Broadly thinking about solutions allows more optimal allocation during problem solving, and brings to life the idea of strategic ambidexterity: a company's ability to explore new problems and business models while exploiting existing ones. Constrained thinking, like silver-bullet thinking, leads to constrained results. For these reasons, I am surprised by the frequency at which I am asked to help organizations find problems for solutions as if creating new problems to solve is the same thing as solving problems.

Big Data

Let's dig into the specific source of all of this awareness. First, let's explore what is often seen as a synergistic relationship between AI and Big Data.

So-called Big Data highlights the fact that more data has been created in the past two years than in the entire previous history of the human race,

---

15. Tim Nudd, "KFC Made a Weird Animatronic Harland Sanders Robot to Take Your Drive-Thru Order," *Adweek*, July 6, 2017, www.adweek.com/creativity/kfc-made-a-weird -animatronic-harland-sanders-robot-to-take-your-drive-thru-order/.

16. For example: https://nypost.com/2018/02/23/kfc-issues-full-page-apology-for-its -chicken-crisis/.

and it continues to grow.[17] In 2020, we will see 1.7 megabytes of new information created every second for every human being on the planet, and our accumulated digital universe will grow to around forty-four zettabytes.[18] Therefore, Big Data underscores an irrefutable fact: there is more data today than yesterday. Data has grown in volume and variety, and almost all objective measures suggest that we are in the midst of "data deluge."[19] AI—specifically machine learning—is useless without data, and there is more data today than at any point in history. Unfortunately, problems are not automagically solved by accumulating data and pouring it into some AI.

Big Data trends have asked us to be more aware of the volume of data. This awareness is often interpreted to suggest that aggregated data has emergent properties. We're told that by collecting more and more of it, more and more value is created. However, data volume is not a useful goal and doesn't produce consistent value for all businesses. When one examines what has been accomplished compared to what was promised, the gap is sobering, because:

1) Big Data doesn't overcome limitations of data.
2) Big Data doesn't tell us anything about which data to collect.
3) Big Data is not a useful goal and doesn't ensure goal alignment.
4) Big Data doesn't relieve us of the responsibility to know where data comes from.
5) Big Data doesn't relieve us of any responsibility of storing data.
6) Big Data does not provide any domain insight to yield a better understanding of the underlying processes that created the data.

---

17. SINTEF, "Big Data, for Better or Worse: 90% of World's Data Generated over Last Two Years," ScienceDaily, May 22, 2013, www.sciencedaily.com/releases/2013/05/130522085217.htm.

18. M. Zwolenski and L. Weatherill, "The Digital Universe: Rich Data and the Increasing Value of the Internet of Things," *Journal of Telecommunications and the Digital Economy* 2, no. 3 (2014): 9, https://doi.org/10.18080/jtde.v2n3.285.

19. C. Anderson, "The End of Theory: The Data Deluge Makes the Scientific Method Obsolete," *Wired*, June 23, 2008, https://www.wired.com/2008/06/pb-theory/.

Big Data and advances in computation are two strong macro-trends that explain, in large part, why machine learning (specifically supervised machine learning) has gained so much momentum in the past decade. After all, solutions that learn from data benefit from more data, and the benefit of more data has been defended by insiders. One of the most famous quotes highlighting the power of data comes from Google's research director, Peter Norvig, who claims that, "We [Google] don't have better algorithms. We just have more data." This quote is usually linked to an article he coauthored with Alon Halevy and Fernando Pereira, "The Unreasonable Effectiveness of Data."[20] While this claim is generally true, it is not *always* true.[21]

It's false in the sense that we must keep in mind that few of us will have what Norvig and Google have. In other words, few of us can truly possess Big Data, because Big Data is a macro-trend. Macro-trends affect businesses differently, if they affect businesses at all.[22] To be sure, macro-trends rarely solve specific problems. Consider that a 2013 *Consumer Science* article by Microsoft researchers presented evidence that suggests that a majority of analytical jobs do not process huge data sets.

Yet many will maintain that *the best data is more data* despite scant evidence to support that claim, contrary evidence that suggests qualitatively different trade-offs, and a complete disregard to the risks of storing data that may ultimately have no value. In practice, Big Data functions more as a type of availability bias, where we think the data we have is the complete data we want and representative of all the data we need. In

---

20. F. Pereira, P. Norvig, and A. Halevy, "The Unreasonable Effectiveness of Data," *IEEE Intelligent Systems* 24, no. 2 (2009): 8–12. Norvig is often misquoted as saying that "All models are wrong, and you don't need them anyway." Read the author's clarifications on how he was misquoted: http://norvig.com/fact-check.html.

21. Xavier Amatriain, "In Machine Learning, What Is Better: More Data or Better Algorithms," KDnuggets, June 2015, www.kdnuggets.com/2015/06/machine-learning-more-data-better-algorithms.html.

22. Many of us hold a notional view that the data is "out there" and just waiting to be collected. This is not always, or even often, the case. At the very least, the data needs to be thoroughly cleaned and wrangled.

reality, the data we have is not the data we often need, with characteristics we never expected or desired. Thus, Big Data oscillates between a map and a territory: seldom the right map and rarely the right territory.[23]

For example, while Big Data is never all data, it is easy to recognize an absolute lack of data. When we have no data, it's easy to understand that we have a problem, especially when leveraging machine learning, which learns inductively from data. Conversely, we may have Big Data that due to some cause lacks all the data. When we lack data due to a particular cause biases often occur, especially if we choose to ignore the absence of data and instead fixate on vanity metrics such as how much data we have. We should focus on discovering what caused us to lack the data we need. By unhinging the evaluation of our data from macro-trends like "Big Data," we quickly create a slippery slope where we can no longer evaluate the quality of what we have or don't have information-wise.

Consider, for example, a lack of historical data in banking and how that missing data could help predict loan defaults. Before a loan could show up in a bank's database, a customer must first have applied for the loan and the bank must have granted the loan based on various known and possibly unknown criteria. Without historical data the current data and algorithm may not provide very good estimates for customer groups who previously would not have applied for that loan or who the bank had screened out prior to the completion of an application. Missing data is not always some random process that is independent of everything else. Big Data offers an opportunity because its volume typically gives us abundant information. But Big Data is not *all* data, so we must always know something about the data and the process that created or failed to create that data.

Furthermore, even all data would not be enough data. That's because no data—regardless of size—can ever be enough data. That is, learning from data is always bound by limits of induction. As a consequence, machine learning can never be truly objective or perfect. For some problems this is acceptable, but machine learning can never be a perfect solution alone regardless of how well it can fit a function or how much data we may have. This does not mean that these solutions are always wrong, rather that they can't always be right.

Moreover, no matter its size, data is neither transparent nor self-evident. Interpretation is still necessary, and interpretation often demands historical, social, and cultural context and certainly produces historical, social, and cultural explanations. There is seldom just one interpretation of data; data never speaks some sort of raw truth to us. Data does not exist before facts and thus data never speaks for itself. We give it a voice—often the wrong one.[24] As renowned British economist Ronald H. Coase famously said, "If you torture the data long enough, it will confess to anything."

The raw data paradigm exists more in our imagination. Consider the idyllic scenario of machine-generated data—as opposed to human-generated data—in cybersecurity. Even when this data is sent directly from a network device to an algorithm (which is often complicated by a host of issues, especially for large, heterogeneous environments), the data is not raw in any substantive sense. Although this data has been generated by machines, these machines were designed by humans who inherited constraints, such as protocols, that predate the very existence of cyberthreats.

While the ubiquity of Big Data technology did drive down costs and integrate disparate data silos, it drove up novelty-based search instead of a problem-based search. What novelty-based searches seek to find are correlations in large data sets that often turn out to be spurious—though even substantive correlations are often distractions from business goals because

---

24. For example, a common expression in Big Data is *let the data speak for itself,* but this should be viewed with skepticism. For example, Lazer et al., concluded that the Google Flu Team engineers overestimated flu trends for more than two years because of what they decided to include, and not include, in their model. Lazer and team dubbed the mistakes "big data hubris." That is, just because web searches can be analyzed does not mean they deserve analysis. D. Lazer, R. Kennedy, G. King, and A. Vespignani, "The Parable of Google Flu: Traps in Big Data Analysis," *Science* 343, no. 6176 (2014): 1203–1205, doi:10.1126 /science.1248506.

needles in haystacks rarely align with previous goals. Insight isn't learning something about data but about problems of importance facilitated by data.

Benchmarks and Contests

Let's go even deeper and explore the role that benchmark data sets and contests play in our awareness. Johns Hopkins University professor Alan Yuille notes how the real world is combinatorially large.[25] What this means is that it will be hard for any data set, no matter how "big," to be representative of real-world complexity.[26]

Computer vision is a large area of research that explores the ability of a solution to identify objects, places, people, writing, and actions in images and videos. Computer vision includes various tasks like image classification, object localization, object detection, semantic segmentation, instance segmentation, and image captioning.[27] Most real-world computer vision

---

25. In mathematics, a combinatorial explosion is the rapid growth of the complexity of a problem due to how the combinatorics of the problem are affected by the input, constraints, and bounds of the problem. Combinatorial explosion is sometimes used to justify the intractability of certain problems.

26. A. L. Yuille and C. Liu, "Deep Nets: What Have They Ever Done for Vision?" *arXiv*, May 2018, arXiv:1805.04025.

27. Image classification refers to the task of assigning an input image on one label from a fixed set of categories. For example, if we have an image of a cat, an image classification model takes a single image and assigns the main object contained within an image based on the input classes or categories. The raw input data is represented as a large three-dimensional array of numbers that represent the values of the pixels in an image. Object localization requires a program to recognize a single instance of an object within an image even if the image contains multiple instances. Object detection is the process of recognizing multiple instances of various objects in images and videos. Object detection is common in security and surveillance. Semantic segmentation is the process of assigning a label to every pixel in an image. Semantic segmentation treats multiple objects of the same class as a single entity. Instance segmentation treats multiple objects of the same class as distinct individual objects. Image captioning is the process of generating textual description of an image. It is not just computer vision but also a natural language-processing problem to generate captions. The data set used in training is in the form of an image-caption pair. Images are commonly fed into a convolutional neural network to extract features and objects, which is then fed into a recurrent neural network that maps the input features and objects to words to describe what is happening in an image.

problems include object frequencies that follow a long tail distribution. That is, we are much more likely to find many occurrences of people, fewer occurrences of coffins, and even fewer of people in coffins. These smaller subcategories are not only infrequent but also vary from problem domain to problem domain. So the subcategories will change not only in occurrence but also in frequency between problems. The smallest subcategories are called edge-cases. Edge-cases are examples that are unexpected.

When we hear of some impressive computer vision result with respect to artificial intelligence, we are actually hearing about the results that yield from the head of some distribution meant to suggest real-world complexity, not the complete distribution and the long tail. What the media would want us to believe is that the 80-percent solution is worth 100 percent of our attention. Yet the last 20 percent of problem solving is usually what takes years to complete and careers to perfect (if they can even be perfected).

Consider the countless edge-cases to enable automated driving on our complex road network. Then consider why companies like Voyage are targeting retirement communities as a deployment environment.[28] Voyage wants to limit the subcategories to enable more consistent performance of their self-driving technology by constraining the problem to a fixed environment with a largely known distribution of objects. Voyage is not solving the complexity, because complexity can't always be solved in an exact way.

Instead, Voyage is reframing the problem in order to find a version of the problem that it can better solve. Ultimately, Voyage seeks to reduce the differences between training a solution and testing that solution, and then to deploy that solution by finding a simplified environment that is consistent throughout these phases. In this way, Voyage may be considered savvy as they seek to constrain the problem to limit edge-cases; though it may be too soon to determine if the problem Voyage is trying to solve—transportation for seniors at retirement communities—has market relevance.

---

28. Oliver Cameron, "Why Retirement Communities Are Perfect for Self-Driving Cars," *Medium*, June 9, 2020, news.voyage.auto/why-retirement-communities-are-perfect-for -self-driving-cars-8bc35edfa804.

Other companies such as Alphabet's Waymo and GM's Cruise don't share Voyage's belief in partially autonomous vehicles or automated driving in constrained environments. Waymo and Cruise are working on at least Level 4, or fully autonomous driving, for their initial commercial offerings. Unlike Voyage, who seeks to create customers in senior communities, Waymo and Cruise are interested in making everyone a customer. To do so, both will have to battle the complexity of the real world represented as edge-cases.

That said, Waymo and Cruise have—by every conceivable measure—Big Data. Waymo has collected data for ten million miles of real-world driving and another ten billion of simulated miles, generating nearly 150 terabytes of data per day, per vehicle, which means they collect about 100 petabytes of data per day.[29] For context, the brain's memory storage is around 2.5 petabytes,[30] and the average person will drive fewer than 700,000 miles in their lifetime.

Even if all storage in the brain was information relating to driving, we can conclude that not only does data hold less information than we think, expect, and often need for problem solving in the real world, but our solutions are not as efficient as we'd hoped. Real-world problems are bigger and more complex due to long tails. The longer the tails, the more data required, and even then it will not be enough. If there is any doubt about the devil in the tails and the inconsistent value of data in the real world, especially on complex problems like autonomous driving, then consider the $16 billion cash burn by self-driving car companies.[31]

---

29. Chris Neiger, "Why It Matters that Waymo Just Passed 10 Billion Simulated Autonomous Miles," *The Motley Fool*, July 12, 2019, www.fool.com/investing/2019/07/12/why -matters-waymo-passed-10b-autonomous-miles.aspx.

30. Paul Reber, "What Is the Memory Capacity of the Human Brain?" *Scientific American*, May 1, 2010, www.scientificamerican.com/article/what-is-the-memory-capacity/.

31. Amir Efrati, "Money Pit: Self-Driving Cars' $16 Billion Cash Burn," *The Information*, February 16, 2020, www.theinformation.com/articles/money-pit-self-driving-cars-16 -billion-cash-burn.

The point is that Big Data is not universal, and even as a subset of all data it fails to capture real-world complexity. Consequently, no amount of complex mathematical or statistical modeling can possibly squeeze more information from data than it contains. Problem solving is rarely just some computational problem of labeling data and training an algorithm. That is why at one end of the spectrum we see companies spend billions as they try to solve the complexity inherent in problems, and on the other end companies attempt to narrow the scope and manage the complexity of real-world problems.

Like Voyage, insiders also narrow the scope of problems and manage complexity rather than trying to solve complexity directly. This is why so many "AI successes" feel toy-like and also explains why our solutions rarely match our awareness of them. It's also why real-world solutions are sloppy in the sense that they will not often work "off the shelf." In practice, an arbitrary solution will not be one magical solution but a number of solutions all working on parts of the problem, and knowledge will come from a variety of places, not just data.

Today, much of our awareness is the result of supervised machine learning and,[32] specifically, of deep learning finding a useful domain of application (more specifically, computer vision). The dramatic performance improvement on computer vision problems like object recognition has been further catalyzed by Big Data and supercharged by benchmark data sets and contests. But the nature of benchmark data sets and contests aren't applied problems as much as they are academic problems, which differ significantly from the real world. While results are impressive, they are academic because they ignore complexity and the countless edge-cases required by a solution in the real world.

---

32. Supervised machine learning learns what are called class boundaries. These boundaries are a kind of bucket that the solution puts new, unseen data into. Today, supervised machine learning does a lot of the heavy lifting in "AI."

The dramatic performance improvements started around 2012, corresponding with University of Toronto professor and Turing Award winner (the highest distinction in computer science) Geoffrey Hinton's winning entry in ImageNet. Hinton, along with Ilya Sutskever and Alex Krizhevsky, also from the University of Toronto, submitted a deep convolutional artificial neural network (CNN) architecture called AlexNet to the ImageNet contest. AlexNet beat the field by a sizable 10.8 percentage points, which was 41 percent better than the next best entry. AlexNet was largely responsible for a resurgence of interest in CNNs for image recognition and triggered the largest rise of supervised machine learning seen today.[33]

Table 3.1 highlights the performance improvements by Hinton over that of previous "shallow" solutions and the performance improvements since then. Deep learning is represented in the unshaded rows. Shallow learning is represented in the shaded row. Shallow learning—also called one-step learning—is so named because estimating parameters of the solution happens in one step by minimizing error on handcrafted features. For example, logistic regression and support vector machines are kinds of shallow learning that fit a function to data in order to minimize error in one step rather than iteratively like artificial neural network architectures.

---

33. AlexNet, originally named *SuperVision*, "almost halved the error rate for recognizing objects in natural images and triggered an overdue paradigm shift in computer vision." AlexNet is one of the most influential papers published in computer vision, cited over eighty thousand times. Alex Krizhevsky, Ilya Sutskever, and Geoffrey E. Hinton, "ImageNet Classification with Deep Convolutional Neural Networks," *Communications of the ACM* 60, no. 6 (2017): 84–90, https://doi.org/10.1145/3065386

Table 3.1 Performance Improvements

Model	Year	Layer Count	Parameter Count	Top-5 Error
Shallow	Before 2012	NA	NA	> 25%
AlexNet	2012	8	61M	16.40%
VGG19	2014	19	144M	7.30%
GoogleNet	2014	22	7M	6.70%
ResNet-152	2015	152	60M	3.60%

ImageNet is the creation of Stanford computer science professor Fei-Fei Li, who sought to build a better data set for testing machine learning algorithms. ImageNet is both a contest and a database containing millions of images and thousands of categories.[34] The ImageNet challenge is large enough to evaluate solutions designed for large-scale computer vision problems, like object detection and image classification.[35] Conceptually, benchmark data sets provide insiders a way to measure success against a known outcome, presumably to make contest results more interpretable. Consequently, benchmark data sets are quite popular, because in theory, benchmarks are more reproducible, less expensive, and more scalable than other options to evaluate solutions.[36]

---

34. The real world does not have thousands of categories—it has millions—and is more geographically diverse than the US and Great Britain. *See* Shreya Shankar et al., "No Classification Without Representation: Assessing Geodiversity Issues in Open Data Sets for the Developing World," *arXiv*, November 2017, arXiv:1711.08536. And the real world contains faces that are not only light-skinned and male. *See* Joy Buolamwini and Timnit Gebru, "Gender Shades: Intersectional Accuracy Disparities in Commercial Gender Classification," *Proceedings of Machine Learning Research* 81 (2018): 77–91.

35. Formally known as ImageNet Large Scale Visual Recognition Challenge or ILSVRC.

36. Kiri L. Wagstaff in "Machine Learning that Matters" (*see* arXiv:1206.4656v1) notes that in practice, direct comparisons often fail because experiments vary in methodology, implementations, or reporting.

Since 2012, contests have proved an effective way to garner a lot of attention very quickly.[37] Winning contestants—those ranked first or high on leaderboards—is an accelerator for many new ventures and yields many new citations. Winners of the contest in years following 2012 have gone on to hold senior positions at Baidu, Google, and Huawei. Matthew Zeiler founded Clarifai on the success of his 2013 ImageNet win, and the company is now backed by $40 million in venture capital funding.[38] In 2014, Google shared the winning title with two Oxford researchers, who were quickly snapped up and added to Google's recently acquired DeepMind.[39]

What is great about ImageNet, as opposed to contests like Kaggle, is that the problem doesn't go away after the contest concludes. ImageNet cares enough about one problem to keep solving it. But although many solutions perform well on benchmarked data sets, they often fail badly on real-world problems. The reason is, in an epistemological sense, benchmark data sets do not reflect reality, resulting in a mismatch between the data used during contests and the data these solutions would otherwise encounter in the real world. This problem is more generally known in the world of machine learning as data shift. Why insiders would choose to create a contest that has little real-world impact explains itself. Insiders

---

37. ResNet was the winner of ILSVRC 2015 and one of the more important victories since 2012. For a review of the 2015 winner, *see* https://towardsdatascience.com/review-res net-winner-of-ilsvrc-2015-image-classification-localization-detection-e39402bfa5d8.

38. Interestingly, Zeiler noticed problems with some of the intermediate layers in AlexNet using deconvolutional networks. Deconvolutional networks effectively map objects to features to pixels instead of pixels to features to objects as is the case with AlexNet, which is a convolutional artificial neural network. Ultimately, deconvolutional networks provided a way to map the feature activity in intermediate layers within an artificial neural network back to the input data and specifically the pixels within an image by attaching a deconvolutional network to each layer of a convolutional network. This effort created a path back to the input data in order to examine AlexNet. Following Zeiler's win, many ImageNet contestants were using deconvolutional networks the following year. Matthew D. Zeiler and Rob Fergus, "Visualizing and Understanding Convolutional Networks," *CoRR* (2013), arXiv:1311.2901.

39. D. Gershgorn, "The Data That Transformed AI Research and Possibly the World," July 26, 2017, https://qz.com/1034972/the-data-that-changed-the-direction-of-ai-research-and -possibly-the-world/.

generally don't want real-world responsibility. In other words, while ImageNet cares about one problem enough to keep solving it, it doesn't care enough about the problem to make it real.

While impressive, contests fuel silver-bullet thinking. The reason why is because benchmark data sets do not reflect reality. Benchmark data sets are not "open" like most real-world problems. Closed-set problem domains ultimately reflect optimal scope and boundedness of an idealized problem without too many ambiguities and inconsistencies. For example, ImageNet does not have distractors or anything that mimics real-world openness. Training and testing comes from known classes where there is little ambiguity among classes, which often occurs when a real problem is poorly defined or poorly conceived.[40]

In fact, if we restrict awareness to reinforcement learning within games (most of which are closed problems) or supervised machine learning within contests, and we measure performance within the constraints of benchmark data sets, then progress toward some silver bullet seems believable.[41] Pundits argue that computers are able to duplicate human performance by seriously simplifying reality. These pundits emphasize the performance of machines in closed domains and point to the inconsistent performance of humans in open domains.[42] They argue that AI is real under conditions

---

40. Problems with overlapping classes are better known to have non-zero Bayes error. The Bayes error is non-zero if the classification labels are not deterministic. That is, there is a non-zero probability of a given instance belonging to more than one class.

41. Julian Togelius argues that the "most important thing for humanity to do right now is to invent true artificial intelligence." *See* J. Togelius, "AI researchers, Video Games Are Your Friends!" Conference Paper, International Joint Conference on Computational Intelligence, November 2017, doi:10.1007/978-3-319-48506-5_1. Togelius makes the argument in support games as the new AI benchmark. Conversely, Rodney Brooks claims that "the world is its own best model." Rodney Brooks, "How to Build Complete Creatures Rather than Isolated Cognitive Simulators," MIT, Artificial Intelligence Laboratory, August 1998.

42. Pundits who argue that computers are able to duplicate human activity are simplifying and distorting reality. They may argue that humans are machines, mere informational processing devices. In AI this is known as the computational metaphor. We should remember that it is just a metaphor. Almost all solutions perform well in closed domains, but nothing, aside from humans, performs well in open domains.

that don't exist in the real world.[43] Philosopher Jaron Lanier has argued that discussions of computers with so-called intelligence are most often expressions of a new religion.[44] At the very least, these kinds of comparisons are disingenuous and subvert our efforts by creating a false imagination.

While machine learning is good at classifying noisy inputs based on known situations in closed-set domains, almost none of machine learning excels at classifying data that is outside of the known problem space represented by data. That's because machine learning learns to fit known data as closely as possible locally, regardless of how it performs outside of these situations. In other words, machine learning is good at interpolation but poor at extrapolation.

To be sure, machine learning cannot generalize to unseen problems in a universal sense. But, even when properly configured, it will struggle to generalize to unseen data. In fact, if we are not careful, we can overfit training data and even fail to interpolate. Today's machine learning is not at all like the informal learning that humans engage in, "making rapid progress in a new domain without having to be surgically altered or purpose-built."[45]

Writer and researcher Luke Oakden-Rayner pejoratively adds that contests like ImageNet combine one part overfitting, meaning solutions are memorizing the training data rather than generalizing the test data while completely ignoring real-world data, and one part multiple hypothesis testing.[46] Multiple hypothesis testing basically means that designers are throwing everything but the kitchen sink at a problem, using many

---

43. Uncertainty, based on the narrow world of games, thus conflating aleatory and epistemic uncertainty. Nassim Taleb, *The Black Swan* (New York: Random House, 2007), 309.

44. Jaron Lanier, *Who Owns the Future?* (New York: Simon & Schuster, 2014), 193, Kindle edition.

45. Rodney Brooks, "The Seven Deadly Sins of Predicting the Future of AI," September 7, 2017, https://rodneybrooks.com/the-seven-deadly-sins-of-predicting-the-future-of-ai/.

46. Luke Oakden-Rayner, "AI Competitions Don't Produce Useful Models," September 23, 2019, lukeoakdenrayner.wordpress.com/2019/09/19/ai-competitions-dont-produce -useful-models/.

solutions that perform very similarly in hope of finding that one solution that wins a competition by fractions of percentage points. These designers often use bigger solutions, more computational power, and complex (hyperparameter) tuning. However, bigger isn't always better. All this is to say that we cannot often tell using contests and benchmark data sets which solutions will work in the real world and which ones won't, once again subverting awareness.[47]

According to Dario Amodei, vice president of research at OpenAI, and Danny Hernandez, research scientist at OpenAI, the amount of computation used to train the biggest solutions has been doubling every 3.4 months since 2012.[48] From 2012 (AlexNet) to 2018 (AlphaGo Zero) the computation required has increased 300,000 times. The architectural search that resulted in AmoebaNets by the Google AutoML team required 450 K40 GPUs for seven days.[49] If done on one GPU, it'd have taken *nine years*. As computer scientist Chip Huyen notes, these massive solutions make for great headlines but not great products.[50] Huyen says that these solutions are "too expensive to train, too big to fit onto consumer devices, and too slow to be useful for users." In other words, insiders are creating bigger and more complex solutions that are inherently opaque, whereas real-world application tends to move in the opposite direction toward small solutions that require less memory and power that are also more interpretable.

---

47. Contests perpetuate the misplaced idea that problem solving occurs by training and testing a machine learning algorithm against examples saved from training. This is often wrong and also a low bar because all the hard work occurs before and after this phase, including noticing a problem, gathering data for training, and, finally, getting a solution to perform well upon deployment.

48. Dario Amodei, "AI and Compute," *OpenAI*, September 2, 2020, openai.com/blog/ai-and-compute/. By comparison, Moore's law had a two-year doubling period. *See* https://newsroom.intel.com/wp-content/uploads/sites/11/2018/05/moores-law-electronics.pdf.

49. Esteban Real, Alok Aggarwal, Yanping Huang, and Quoc V. Le, "Regularized Evolution for Image Classifier Architecture Search," *arXiv*, February 5, 2018, arXiv:1802.01548.

50. Chip Huyen makes this salient point as well. *See* Chip Huyen, "Machine Learning Interviews," *GitHub*, November 24, 2019, github.com/chiphuyen/machine-learning-systems-design/blob/master/build/build1/consolidated.pdf.

It would appear that some of the world's greatest minds are working on the margins of a "solved" contest with benchmark data sets rather than real-world problems with social and economic purpose. In fact, contests seem to be causing, or at least reflecting, a growing gap between scholars (i.e., insiders) and practitioners (i.e., outsiders). Insiders are becoming strangely radicalized around contests and benchmark data sets, and they hyperfocus on marginal or incremental improvements.[51] For example, one survey of 152 papers at the 2011 International Conference of Machine Learning revealed that just 1 of 148 papers interpreted results in the context of a real-world problem.[52] Although there is value for insiders in exploring empirical and theoretical boundaries while they figure out how to make solutions better, smaller, and faster, there is often little value for business to ignore real-world problems. Only trust those with dirt on their hands. All others are almost certainly wrong.

In defense of ImageNet, it is not the only academic repository that lacks complexity. An analysis of the popular University of California, Irvine (UCI), Machine Learning Repository exposed that the majority of UCI problems are easy to solve, with only 3 percent deemed challenging for the supervised machine learning classifiers tested.[53] This is not a revelation, either. The late Carnegie Mellon University professor, Jaime Carbonell, wrote in 1992 that "the standard Irvine data sets are used to determine percent accuracy of concept classification, without regard to performance on a larger external task."[54]

51. Yoshua Bengio, "Time to Rethink the Publication Process in Machine Learning," *Yoshua Bengio*, March 3, 2020, yoshuabengio.org/2020/02/26/time-to-rethink-the-publication-process-in-machine-learning/.

52. Kiri Wagstaff, "Machine Learning that Matters," Proceedings of the Twenty-Ninth International Conference on Machine Learning (ICML), *arXiv*, June 2012, arXiv:1206.4656, 529–36, http://arxiv.org/abs/1206.4656.

53. Núria Macià and Ester Bernadó-Mansilla, "Towards UCI+: A Mindful Repository Design," *Information Sciences* 261 (March 2014): 237–62.

54. Jaime Carbonell, "Machine Learning: A Maturing Field," *Machine Learning* 9 (1992): 5–7.

Of course, some researchers do care about making problems less perfect and, ultimately, more real. Researchers recently curated 7,500 natural adversarial examples and released them in an ImageNet classifier test data set that they call ImageNet-A.[55] Additionally, insiders have created more complex artificial data sets to increase the diversity of the UCI repository. The UCI+ repository is not a substitute for real problems, but it has at least been more carefully designed to span a representative space of real-world problems.

As previously mentioned, Kaggle is also a competition. As opposed to ImageNet, it invites applied data scientists to compete to produce the best solutions for real business problems. Unfortunately, by solving the hardest problems for contestants, Kaggle removes all aspects of what makes a real problem a real business problem.[56] For example, Kaggle contestants don't have to worry about getting a solution to run in real time on live data. They also don't have to worry at all about concept drift, data drift, and any responsibility to customers. Moreover, Kaggle hides all of the engineering challenges inherent in real problems, such as cleaning and collecting data and developing data pipelines that scale. Kaggle also hides all of the business aspects of business problems by choosing a problem, defining a problem, and ultimately selecting the right problem from other competing problems. And, unless we believe that prediction accuracy is the only metric that is right for business, Kaggle rarely adopts metrics that are right for businesses with problem-specific information.

While performance increases have been noteworthy, professor and former chairman of the computer science department at UCLA Adnan Darwiche says, "it highlights problems and thresholds more than it highlights

---

55. Dan Hendrycks et al., "Natural Adversarial Examples," January 2020, Version 3, arXiv: 1907.07174. Dan Hendrycks and Thomas Dietterich also made a variant of the ImageNet benchmark for image classifier robustness. *See* Dan Hendrycks and Thomas Dietterich, "Benchmarking Neural Network Robustness to Common Corruptions and Perturbations," *arXiv* e-prints, March 2019, arXiv:1903.12261.

56. Julia Evans, "Machine Learning Isn't Kaggle Competitions," *Julia Evans* (blog), June 19 2014, jvns.ca/blog/2014/06/19/machine-learning-isnt-kaggle-competitions/.

technology."[57] In other words, solutions having found some useful problems to solve, which means we have largely discovered success of its own choosing. All in all, these factors result in something more akin to the informal fallacy of the infamous Texas sharpshooter rather than the creation of a silver bullet. The Texas sharpshooter fires randomly at a barn door and then paints targets around the bullet holes, creating a false impression of being an excellent marksman. The sharpshooter symbolizes the dangers of post-hoc theorizing or finding your problem in a solution.

Darwiche's statement underscores something many of us intuitively understand, which is the idiom of a solution in search of a problem. The assertion that a "solution is in search of a problem" is sometimes pejoratively made when problem solving has been done backward. The preferred process is to have "a problem in search of a solution," where we identify a problem and work to discover the solution. Once you decide that you don't want a solution in search of a problem, you must notice problems. The key to business success is noticing the right problems and solving them the right way. Our objective is to drive business outcomes, and we cannot achieve it by monitoring contests.

We may start asking ourselves basic questions such as, Can nature be quantified by supervised machine learning, on a data set like ImageNet, and within the constraints of a contest? Physicist Richard Feynman noted that for a successful solution, science must take precedence over public relations, as nature cannot be fooled.[58] At the very least, we must accept that success of its own choosing will not always translate to success of your choosing, mainly because few problems are equivalent to other problems.

---

57. A. Darwiche, "Human-Level Intelligence or Animal-Like Abilities?" Technical report, Department of Computer Science, University of California, Los Angeles, July 2017, arXiv: 1707.04327.

58. "Report of the Presidential Commission on the Space Shuttle Challenger Accident," NASA, June 6, 1986, Appendix F: Personal Observations on the Reliability of the Shuttle, https://science.ksc.nasa.gov/shuttle/missions/51-l/docs/rogers-commission/Appendix -F.txt.

No Free Cake

The unlikelihood of a silver bullet is expressed in the No Free Lunch Theorems (NFLT).[59] The NFLT explains why you can't have your cake and eat it too. What the NFLT and the cake proverb tell us is that no solution is better than all others, over all problems. Although some algorithms are better than others on some subset of problems, they are not better than all others on all other problems. Today, only human performance is better than all others "in the extremely broad class of complex and difficult learning problems that tend to appear in our world."[60]

A terse summary of the theorems is that broad performance in machine learning is found to be low when averaged over all types of problems.[61] The reason is that the learning bias of a machine learning algorithm is the set of assumptions that a solution uses to predict outputs given inputs that it has not yet encountered.[62] In fact, the central goal of supervised machine learning is guided by the question, Will my learning algorithm generalize from a source domain (or training and test environment) to a target domain (my real-world deployment)? While there are many ways to solve problems that fall well outside of the machine learning paradigm (especially supervised machine learning), the problem of generalizing to unseen data is much harder to accomplish than our awareness of AI suggests. That is, without any additional information about a problem, we cannot exactly

---

59. Martin Sewell, "No Free Lunch Theorems," www.no-free-lunch.org/.

60. Christophe Giraud-Carrier and Foster Provost, "Toward a justification of meta-learning: Is the no free lunch theorem a show-stopper?" January 2005, Proceedings of the ICML—2005 Workshop on Meta-learning.

61. Interestingly, Shane Legg and Marcus Hutter (as already discussed) roughly define artificial intelligence as the average performance of an agent on all possible problems where David Wolpert and William Macready find that broad performance in machine learning is low when averaged over all types of problems. In other words, AI is defined by some as a free lunch.

    S. Legg and M. Hutter, "Universal Intelligence: A Definition of Machine Intelligence," *Minds and Machines* 17, no. 4 (2007): 391–444.

62. This relationship between inputs and outputs would not necessarily exist with general intelligence.

know if a solution will perform well because unseen situations might have arbitrary values.[63]

Therefore, the conventional interpretation of the NFLT is that a silver bullet is impossible, and the only way one problem-solving strategy can outperform another is if the solution is specialized to the specific problem under consideration. Said differently, problem-specific knowledge is required to design a successful solution, and the NFLT underscores why it is necessary to understand the relationship between a problem and a solution.

The StatLog project is one of the largest comparisons introduced among algorithms on a large number of data sets.[64] Using statistical analysis, solution performance is measured on both the percentage of correct classifications and computational complexity. Consistent with the basic idea of the NFLT, the project finds that no algorithm is uniformly accurate over the data sets studied. This makes sense. After all, there is a reason that machine learning has thousands of solutions and countless derivations and combinations.

In any case, it's quite analytical to claim that one algorithm is better than another, because you're not saying that one is superior to all others, much less that one is superior to any others. To say that a solution matches a problem better than another is not to say that either is specialized to a problem or specialized at all. When solutions aren't specialized to a problem, any given solution may outperform another on a problem, so you may end up with the best of the worst solutions. When a solution is in fact specialized for a problem, that solution will perform worse than all others on any other problem, because that solution is so specialized for that problem. As professor at the Santa Fe Institute David Wolpert and chief scientist at

---

63. As an aside, the arbitrariness between our training and test environments (i.e., source domain) and deployment (i.e., target domain) is a wonderful reflection of the complexity of the underlying problem and reflects what may be called a generalization gap. The larger the gap the harder the problem.

64. C. Henery, "StatLog: An Evaluation of Machine Learning and Statistical Algorithms," in *Computational Statistics*, eds. Y. Dodge and J. Whittaker (Heidelberg, Germany: Physica-Verlag HD, 1992), 157–162.

Sanctuary AI William Macready put it: if algorithm A outperforms algorithm B, then loosely speaking there must exist exactly as many solutions where B outperforms A.[65]

The same can be said about a company that has a specialized solution. That company has been specialized for their problem (or part of the problem), their customer, and their market. Therefore, that company will tend to underperform with other customers and in other markets. Ultimately, a solution's performance and a company's performance are determined by how aligned they are to the problem they solve, not how aligned some solution is to the external, futuristic, kaleidoscopic goals of artificial intelligence.

Consider cybersecurity and the nearly two hundred purported AI cybersecurity companies.[66] There is no uniform performance among these solutions or these companies. These companies are working on different parts of the problem, and all are attempting to solve their part differently; mainly because they are all creating different customers who are all ultimately interested in specific parts of the problem they value most. Even those companies focusing on the same parts of the problem are not solving those parts the same way. There isn't one solution, and often there isn't one problem to solve or one interpretation of the same problem. And there isn't necessarily just one customer to create.

Consistent with the NFLT, any one machine learning algorithm is not always reliable on different problems, the same problem with different data, or the same problem at different times. Most all machine learning

---

65. Roughly speaking, Wolpert and Macready precisely wrote: "If algorithm A outperforms algorithm B on some cost functions, then loosely speaking there must exist exactly as many other functions where B outperforms A." Wolpert and Macready wrote a series of papers on the subject from 1995 to 2005 including: "No Free Lunch Theorems for Search," Working paper, Santa Fe Institute, 1995, https://www.santafe.edu/research/results/working-papers/no-free-lunch-theorems-for-search; Technical Report SFI-TR-95-02-010, Santa Fe Institute, 1997; "No Free Lunch Theorems for Optimization," *IEEE Transactions on Evolutionary Computation* 1, no. 1 (April 1997): 67–82, doi:10.1109/4235.585893; "Coevolutionary Free Lunches," *IEEE Transactions on Evolutionary Computation* 9, no. 6 (December 2005): 721–35, doi:10.1109/TEVC.2005.856205.

66. Global Artificial Intelligence Database Asgard 2018 (n=186), https://asgard.vc/global-ai/.

solutions need to be retrained, reengineered, or retuned as a problem changes or drifts. Solutions often require bespoke design and targeted features for targeted training, which does not always or even frequently generalize to other problems, the same problem, or other data sets and time periods, and certainly not all other problems at all other times.

Moreover, a solution trained on problem A will be unable to perform well at problem B even if they are related. In other words, the experience gained by training a machine learning solution to predict something like cats does not help it to find anomalies in MRI scans. Performance in terms of inference is even inconsistent using so-called transfer learning. If you have heard of transfer learning, then you probably know that it's an interesting area of research that gets quite a bit of buzz in the press.

Transfer learning happens to be an important area of research for insiders, and this should come as no surprise. As we've discussed solving *a* problem is not the domain of insiders as they seek to solve everything. Insiders generally care more about universal performance than about performance for a specific solution conditioned on a problem. In the absence of universal performance, they favor faster training between problems. That is, insiders generally care more about a solution that can be trained to behave more generally—even if it ultimately serves few, if anyone—than to help many by deploying a narrow solution.

Conceptually, transfer learning makes it possible to train an algorithm and transfer experience to a closely (emphasis on *closely*) related task, provided additional labeled data is available for that second-order problem or task. How transfer learning works is that a solution is developed for problem A then reused as the starting point for solving problem B. Intuitively, this happens because the features learned by the initial solution also capture related data structures that are useful for both problems.[67] Though, in practice, transfer learning requires comprehension of not one problem but

---

67. Transfer learning is a type of strong prior. A prior is assessed before making reference to certain relevant observations and, as the name indicates, *prior* to the problem or the task to perform.

two, problem *A* and problem *B*. Transfer learning has not rendered problem comprehension obsolete. In fact, the opposite is true. Transfer learning is only successful when you understand how problems share similarities with other problems.

There are many empirical areas of study that have investigated the NFLT. One such subfield of machine learning that explores the relationships between the performance of a solution and the underlying complexity of a problem (using characteristics of a problem as computed on the data) is called meta-learning (MtL).[68] In this area of research meta-learning is used is to support the recommendation and selection of various machine learning algorithms as well as their configurations.[69]

An initiative that demonstrates the value of meta-learning is OpenML, which is an online research platform that supports a standard characterization of data sets to improve the recommendation and selection of machine learning solutions.[70] The goal of OpenML is to help a solution determine beforehand if it is right for a problem. AutoKeras, AutoWeka, TPOT, Sage-Maker, AutoML, and Neural Architecture Search are all tools that attempt to automate the selection of solutions, thereby streamlining the trial-and-error approach of trying several solutions to achieve a satisfactory fit for a particular problem. These efforts seek to help data scientists become more productive and efficient by delegating solution selection to other solutions. That said, tools that automate this selection process are still buggy and will

---

68. E. Leyva, A. Gonzalez, and R. Perez, "A Set of Complexity Measures Designed for Applying Meta-Learning to Instance Selection," *IEEE Transactions on Knowledge and Data Engineering* 27, no. 2 (2015): 354–67.

69. Such recommendations are based on so-called metadata, consisting of performance evaluations and characterizations of data sets. These characterizations—called meta-features—describe properties of the data that are at least descriptive and often predictive for performance of a solution trained on them.

70. Professor Adriano Rivolli presents a Meta-Feature Extractor (MFE) tool to compute many of these same meta-features present in OpenML. Publicly available as a package in Python and R, MFE offers a flexible and stand-alone implementation of meta-features for MtL experiments. *See* A. Rivolli et al., "Characterizing Classification Datasets: A Study of Meta-Features for Meta-Learning," *arXiv*, August 2018, arXiv:1808.10406, https://arxiv .org/abs/1808.10406.

not replace data scientists or project managers because many real-world problems are too complex for the selection of canned solutions.

In other words, tools like AutoML do not solve the NFLT, nor do they replace problem solving or bespoke design using problem-specific knowledge. Picking the right solution is still only useful once we have found the right problem, and we cannot determine the right problem by looking at what problems canned solutions are already solving. AutoML and MtL will never find the right problems to solve. In fact, that would be ASI. While meta-learning helps us understand problem complexity as captured by the input data, they do not tell you anything about the right problem.

This isn't to say that machine learning is weak. In fact, it has been adopted in a wide range of domains where it shows superiority over other solutions such as traditional software development where you must know every detail about a problem or expert systems. Expert systems, or so-called knowledge-based systems, use databases of expert knowledge to offer advice using rules, or so-called inference engines, in areas such as medical diagnosis and stock trading. Expert systems store and manipulate knowledge to interpret information in a useful way. They are called expert as they capture specialized knowledge and not because they possess some kind of general aptitude or surpass the ability of humans.

The reason why machine learning has been adopted in a wide range of domains is that it relaxes many of the requirements of problem solving. Specifically, the requirement of having to know every detail about a problem, which is a requirement for traditional software development and expert systems. The fact that aspects of almost every problem could be learned is an especially useful remedy to alleviate Polanyi's paradox, which states "we know more than we can tell." Polanyi's paradox is named in honor of the philosopher and polymath Michael Polanyi. The paradox explains how some problems are too hard to tell or require too much to tell.[71] There may be no other development like machine learning, either in

71. To be fair to Monica Anderson, her assertion that intelligence *is what is not,* is intelligence is similar to Polanyi's paradox. Polanyi's paradox highlights why the tasks requiring

technology or management, that by itself promises this type of efficiency gain in problem solving. Unfortunately, it does not replace any of the previously mentioned technologies. What's more, is that NFLT reminds us that problems matter to our solutions.[72] We must not forget that problems matter most to our customers.

What Our Solutions Look Like . . .

Awareness and silver-bullet thinking is also influenced by what our solutions look like and the way in which we value aesthetics. This is especially true for those who wish to simply call deep learning "AI," in a desire to have AI done already. As discussed previously, deep learning is a label that describes a type of architecture, a type of methodology, and a family of machine learning algorithms that roughly mimic neurobiological structures.[73] Of course, "roughly mimic" because the anthropomorphic credentials of artificial neural networks are quite overstated.[74] Nonetheless, deep learning is already present in our lives whether we know it or not.[75]

For example, researchers at New York Eye and Ear Infirmary of Mount Sinai have developed deep learning solutions that can detect age-related

---

understanding and common sense have proved most difficult to automate. The reason is that these problems require skills that we understand only tacitly.

72. A similar argument made by Fred Brooks in "No Silver Bullet," http://worrydream .com/refs/Brooks-NoSilverBullet.pdf.

73. Deep Learning is based on very early models of the neuron. Better models exist but are usually used more in theoretical neuroscience.

74. David Watson, "The Rhetoric and Reality of Anthropomorphism in Artificial Intelligence," *Minds and Machines* 29, no. 3 (2019): 417–40.
    Although outside the scope of this book, Yoshua Bengio explains in a 2015 paper that deep learning offers a machine learning answer (i.e., computational answer), but it is not biologically plausible—a point often misconstrued by many. *See* Yoshua Bengio et al., "Towards Biologically Plausible Deep Learning," *arXiv*, February 2015, arXiv:1502.04156, http://arxiv.org/abs/1502.04156.

75. Deep learning software can be found in various places including Neural Designer, H2O.ai, DeepLearningKit, Microsoft Cognitive Toolkit, GoogleNet, ConvNetJS, Torch, Deeplearning4j, Gensim, Apache SINGA, Caffe, Theano, Keras (capable of running on top of TensorFlow and Theano) ND4J, and NXNet.

macular degeneration, a leading cause of blindness in the United States.[76] The ability of deep learning to classify images of skin lesions as benign lesions or malignant skin cancers achieves the accuracy of board-certified dermatologists.[77] In other words, DL is good—often great—overlapping with narrow cognitive functions like perception and spatial reasoning. Consequently, DL is both familiar—mimicking neurobiological structures that are distinctly familiar to humans—and novel in terms of performance on a small variety of tasks.[78] These opposing mental evaluations of being both familiar and novel in terms of performance makes "deep learning" especially potent as a label and consequently as a brand.

While DL can do a lot, there is a general feeling that deep learning can *do it all* due to the fact that its algorithm shares design features with the brain. It should come as no surprise that we find the biologically inspired design aesthetically appealing given our pride and self-love. But aesthetics distort reality. Quinn Norton created a humorous image that shows a fictitious area of the brain that shuts off when seeing pretty pictures of the brain.[79] Norton is spoofing the tendency to treat experimental results as more scientific when they are accompanied by images of the brain. To be sure, pictures speak more directly to us than words. You should be particularly suspicious of anyone who shows you images of a brain when talking about their solution.[80]

Another study found that adding a line of meaningless "neurobabble" (the neuroscience equivalent of psychobabble) to the explanation of

76. Mount Sinai Health System, "Artificial Intelligence Algorithm Can Rapidly Detect Severity of Common Blinding Eye Disease," *Newswise*, May 12, 2020, https://www.newswise.com/articles/artificial-intelligence-algorithm-can-rapidly-detect-severity-of-common-blinding-eye-disease.

77. A. Esteva et al., "Dermatologist-Level Classification of Skin Cancer with Deep Neural Networks," *Nature* 542 (2017): 115–18, https://doi.org/10.1038/nature21056.

78. Though not so good at motor commands. *See* https://www.youtube.com/watch?v=g0TaYhjpOfo.

79. Quinn Norton, https://www.flickr.com/photos/quinn/4252155172/.

80. This is also why no images of neural networks—natural or artificial—are shown in this book.

a scientific result led people to find that explanation more compelling.[81] It seems just mentioning the brain and biologically inspired design leads people to think a solution is better than it is, even though such neurobabble adds nothing of logical or descriptive value to a solution. Neurobabble succeeds because it plays on the illusion of knowledge. We tend to believe more in something that shares design features with our own biology and may attribute qualities to a solution that the solution doesn't have.

This is *not* to suggest that neurobiological architectures (i.e., artificial neural networks and deep learning) are not great at pattern recognition. Nor does this suggest that neuroscience, psychology, and artificial intelligence don't or shouldn't cross-fertilize one another to spur progress within each field. Merely that in practice we gravitate to solutions that seem more familiar based on their looks. Consequently, we think we know more about a solution than we do, and, more perniciously, we begin to think the solution knows more than it does.

More importantly, commitments based on a solution's aesthetics are commitments based on looks rather than a commitment to a problem being solved. Aesthetics are never a good commitment because any solution—even the ugliest or the simplest—that satisfies a problem can be considered a good solution. Ultimately, elegance and beauty are not directly related to usefulness, because they are attitudes. Specifically, they are insider attitudes, and attitudes have inconsistent practical value. For insiders, solutions that are beautiful are desirable because beauty often mimics something deep and profound in nature. And if such a design works in nature, it is anticipated to work elsewhere. In other words, insiders believe that beauty is truth and truth is beauty.

However, the distinction between beauty and utility is fundamental. Elegance provides pleasure whereas utility provides function. Consider that when Carlos Guestrin, professor of machine learning at the University

---

81. S. O. Lilienfeld et al., "Neurohype: A Field Guide to Exaggerated Brain-Based Claims," in *The Routledge Handbook of Applied Ethics*, eds. L. S. M. Johnson and K. S. Rommelfanger (New York: Routledge/Taylor & Francis Group, 2018), 241–61.

of Washington, recently answered questions about his favorite algorithmic solution he chose the perceptron algorithm.[82] Guestrin said, "perceptron algorithm is one of the most elegant pieces of math I've seen." Guestrin nominated boosting as the most useful algorithm—specifically, boosted decision trees.

The perceptron is a single-layer artificial neural network. Developed in the 1950s by Frank Rosenblatt, the simple perceptron algorithm can be viewed as the foundation for some of the most successful machine learning solutions today, including support vector machines and artificial neural networks. Rosenblatt's book, *Principles of Neurodynamics: Perceptrons and the Theory of Brain Mechanisms*, published in 1962, explains his approach to modeling the nervous system in such a way that it could be applied to data.[83] His research raised a lot of enthusiasm and inspiration for neurobiological architectures.

A boosted decision tree is used for both classification and regression problems. Regression and classification are types of supervised machine learning. In general, decision trees are in the form of if-then-else statements, and tree structures can represent deep, complex rules.

The idea of "boosting" is that a weak solution can be modified to become a stronger solution. Boosting uses small models, filtering easy observations that a weak model can handle, and focuses on developing smaller, weaker models to handle remaining difficult observations. The goal is to use weak models several times to get a succession of solutions, each one refocused on the parts of a problem that the previous solutions found difficult to predict or classify. The prediction after "boosting" is a weighted average of the predictions by multiple weak models.

Boosting is often used to reduce prediction error and is a good hedge against the rather serious problem of overfitting. Overfitting is a modeling

82. Quora, "These Are the Most Elegant, Useful Algorithms In Machine Learning," *Forbes*, October 19, 2016, www.forbes.com/sites/quora/2016/10/19/these-are-the-most -elegant-useful-algorithms-in-machine-learning/.

83. F. Rosenblatt, *Principles of Neurodynamics: Perceptrons and the Theory of Brain Mechanisms* (Washington DC: Spartan Books, 1962).

error that occurs when a function is too closely fit to a limited set of data points, particularly data points used for training a machine learning algorithm. Overfitting generally takes the form of making an overly complex solution to spuriously explain idiosyncrasies in the data that subverts generalization. The term *generalization* is borrowed from psychology. In machine learning terms, generalization represents a type of underfitting and refers to a model's ability to adapt properly to new, previously unseen data drawn from a similar distribution as the one used to train the model. In the real world, the goal is to balance overfitting and underfitting to obtain high performance and generalization. If a model doesn't overfit, then it is always a little wrong. But it's wrong for the sake of generalization.

Doctoral candidate at the Oxford Internet Institute, David Watson, writes in an article titled, "The Rhetoric and Reality of Anthropomorphism in Artificial Intelligence," how the applicability of boosting should come as no surprise to anyone familiar with the so-called wisdom of the crowds.[84] Journalist James Surowiecki presents his theory of crowds in his book, *The Wisdom of Crowds*, saying aggregation plays an important role in wisdom. The idea is that large groups of people are collectively smarter than individual experts. In our case, boosting aggregates many smaller solutions together and is therefore "smarter" than an individual solution.

Watson also highlights how the sequential nature of boosting bears some striking similarities to a cognitive process called predictive coding. According to this theory, human perception is a dynamic inference problem in which the brain is constantly attempting to classify objects and update predictions based on new information.

Without going into too much more detail, boosting refers to a type of meta-learning called ensemble learning. In addition to boosting, ensemble learning includes bagging, stacking, and cascading. Instead of learning sequential models, as boosting does, bagging learns to fit various models

---

84. Watson, "The Rhetoric and Reality," 417–40.

independently to one another and is commonly used to reduce the variance in a model.[85]

As a matter of fact, Cybraics has been researching meta-learning
conceptually—as well as developing meta-algorithms practically—for cybersecurity, building upon years of research the Cybraics data science team
conducted at the Defense Advanced Research Projects Agency (DARPA).
Cybraics' solution uses a kind of multitask meta-algorithm for distributed
learning over the whole of the cybersecurity problem. Meta-algorithms
are an important development in iterative and adaptive computations that
show problem solving as dispersed and often continuous.[86] Cybraics prefers
both meta-learning and meta-algorithms over the research and development of general-purpose learning algorithms. The general idea of combining information from multiple sources and creating a strong solution by
combining many solutions can be applied broadly.

---

85. Models with high variance pay a lot of attention to training data. They don't generalize on the data they haven't seen before, which is bad for real-world solutions. A very
popular example of bagging is Random Forest developed by statistician Leo Breiman. See
L. Breiman, "Random Forests," *Machine Learning* 45 (2001): 5–32. https://doi.org/10.1023
/A:1010933404324.

Random Forests have impacted many academic disciplines including genomics,
econometrics, and computational linguistics. Stacking is used to increase the prediction accuracy of a model. The idea of ensembles is so popular (*see* David Warde-Farley
et al., "An Empirical Analysis of Dropout in Piecewise Linear Networks," *arXiv*, arXiv:
1312.61972013, http://arxiv.org/abs/1312.6197; K. Hara et al., "Analysis of Dropout Learning Regarded as Ensemble Learning," *arXiv*, doi:10.1007/978-3-319-44781-0_9; *and*
Wojciech Zaremba, Ilya Sutskever, and Oriol Vinyals, "Recurrent Neural Network Regu-
larization," *arXiv*, arXiv:1409.23292014, http://arxiv.org/abs/1409.2329) that the process is
used for combining neglected hidden units in artificial neural networks: something called
dropout learning. *See* Kazuyuki Hara, Daisuke Saitoh, and Hayaru Shouno, "Analysis of
Dropout Learning Regarded as Ensemble Learning," *arXiv*, June 2017, arXiv:1706.06859,
http://arxiv.org/abs/1706.06859. Dropout learning randomly drops units in a neural
network during training to prevent co-adaptation (*see* G. E. Hinton et al., "Improving
Neural Networks by Preventing Co-adaptation of Feature Detectors," *arXiv*, July 2012,
arXiv:1207.0580, http://arxiv.org/abs/1207.0580)—which is a type of overfitting in neural
networks—and prioritizes generalization over accuracy.

86. NFLT is true even for deep learning, where we must find the right architecture to best
fit our problem and conduct extensive hyperparameter tuning to fit data. The point is not
that finding a satisfactory solution is impossible, merely that there is no silver bullet for
all problems.

All this aside, we ought to be in vigorous agreement with Guestrin—at least in that ensemble learning and boosting are useful. We may ask ourselves, however, why the most useful solution is not also the most elegant even with what seems like some elegance because of the wisdom of the crowds and some cognitive plausibility like predictive coding.

The perceptron is elegant, but we must recognize that elegance does not directly relate to usefulness. In fact, in 1969, a rather damning critique by Marvin Minsky and Seymour Papert on perceptrons stagnated research and demonstrated that the perceptron was not useful. Their paper, titled *Perceptrons: An Introduction to Computational Geometry*, showed that perceptrons could not solve the classic example of a linearly inseparable pattern called the logical exclusive-OR (XOR) function.[87] Although a single-layer perceptron network is able to categorize data into two classes as the linear threshold function defines their linear separability, the two classes *must* be linearly separable for the perceptron network to function correctly. Indeed, this is the main limitation of a single-layer perceptron network, which led to the invention (or, depending on your philosophical bent, discovery) of today's multilayer artificial neural networks.

The distinction between beauty and utility is fundamental. Elegance provides pleasure, whereas utility provides function. This creates what I call an elegance syndrome—where we seek beauty rather than utility— which is ill-advised for business. We ought to always challenge notions of the most beautiful solution in favor of, or at least consideration of, the best, worse, easiest, simplest, smallest, cheapest, most interpretable, or easiest to debug. It doesn't matter how we deliver value, just that value is delivered. Beauty is rarely valuable because it's highly unlikely that your customers will ever look at your algorithm(s). If value happens, customers will notice, and if value does not happen, they will not care how it was not delivered.

For practical context, decision trees and random forests are the second most commonly used solutions behind none other than the most boring

87. Marvin Minsky and Seymour Papert, *Perceptrons: An Introduction to Computational Geometry* (Cambridge, MA: MIT Press, 1969).

solutions, linear and logistic regression.[88] Although these solutions are not elegant or as cognitively plausible as deep learning, cannot always fit the same complex functions to data, and do not get the same press coverage, they do produce useful results for many problems, are easy and inexpensive to use, and are easy to interpret. In fact, many real-world solutions are built with these methods as part of the solution stack that would certainly be rejected by misplaced attitudes about the nature of solutions, while moderately performing middle-of-the-road solutions using more fashionable designs are accepted. Avoid some odd competition for a status badge that cannot be won. There is more than enough competition in markets.

Fluency Heuristics and Vanity Metrics

Finally, awareness is influenced by fluency heuristics. A fluency heuristic hypothesizes that subjects evaluate proposals more highly based on attributes that are easier to evaluate, rather than by those that are important. In other words, we tend to have more awareness of a solution if it is easier to be aware of it. Consider Big Data, which was developed in the marketing sector to allow businesses to focus their energies and to critique various technologies. However, the evaluation of Big Data is mainly an evaluation of the volume of data, not value, because it's easier to be aware of volume.

For example, we can easily claim that 1PB of data is 1,024 times larger and, therefore, 1,024 times better than a 1TB of data. Although data often gives us more information, the fallacy is that more data provides proportionately more information or new information at all.[89] Some people do not need any more data, whereas others may need a lot more; it depends on the goal and the problem.

---

88. "Kaggle's State of Data Science and Machine Learning 2019," https://www.kaggle.com/kaggle-survey-2019.

89. For example, one million images of hot dogs may be no more valuable than one thousand images of hot dogs.

What is clear is that what we don't need is an arbitrary amount of data. We need the right data, at the right time, to solve the right problems. As we cannot store more data than we have, storage is always a function of utility. The question is really what data should we collect and keep. You may believe that you lack the necessary Big Data technology to manage something like unstructured data, but what many organizations lack is an ability to collect the right data in the first place. Having data that does not align with a problem or exceeds a goal is not more valuable; it's just more expensive.

Let's return to the role of fluency heuristics in AI and specifically deep learning. Literary experts (who are often self-assigned "futurists") will suggest that knowing the distinction between deep learning and artificial neural networks is what businesses need to know to achieve business gains.[90] Ah. But this is blatantly wrong. There is no connection between the name of a solution and business-related outcomes. Nor are business gains and deeper solutions analytically identical. In fact, anyone in business knows that business gains cannot be relegated to a single variable, much less one as abstract as the depth of an artificial neural network. Most of us use more variables when selecting business attire than someone not part of your business would have you believe about your own business success.

"Literary experts" are neither insider nor outsider. They are not practicing businesspeople or machine learning experts. Yet, literary experts are commonly found discussing the topic of artificial intelligence. For the literary expert, the greatest advancement in AI is the air of good fortune that accompanies the fact that they can now count solutions by a vanity metric like hidden layers instead of measuring solution performance against an actual problem with problem-specific information, which has at least a

---

90. Bernard Marr (@BernardMarr), "Do You Understand The Difference Between #Deep #Learning And #Neural #Networks? #Bigdata and #AI have brought many advantages to #businesses in recent years. But with these #advances comes a raft of new #terminology that we all have to get to grips with," Twitter, January 20, 2019, https://twitter.com /BernardMarr/status/1086998222982254592; see also, https://www.linkedin.com/pulse /do-you-understand-difference-between-deep-learning-neural-marr/.

chance of impacting business. For literary experts, fluency heuristics are gold as the "expert" can solution-guess and—incorrectly—assert that a few extra hidden layers, plus back-propagation, will automagically solve any problem without talking about the problem. In other words, the literary expert is master at seeing the little picture. To them, with more hidden layers every problem can be converted into a larger electricity bill.

Literary experts espouse solution-centric thinking that often explains its own existence. It exists because the alternative is much harder to produce. These strategies implicitly highlight how hard it is to otherwise find the right problems to solve. To find, acquire, or create the right data, and to commit to solving a problem and change the behaviors of users and customers, always requires more commitment than solution guessing. Ultimately, solution-centric thinking leaves everyone safe from any responsibility for action.

What is lost in all of this noise is that business value is never captured in abstract concepts of deeper and more complex solutions.[91] In fact, there are no real advantages for business to know the difference between deep learning and artificial neural networks, because the goal of business is not to have deeper, more complex solutions. Certainly, deeper (pun intended) comprehension of problems is more valuable than accumulating and counting hidden layers. After all, any fool can make a solution bigger and more complex, but it is rarely a good idea.

The depth of an artificial neural network denotes the number of levels or layers. Typically, in deep learning there is one input layer, one output layer, and at least one (but often more) hidden layers. Hidden layers are called hidden because they are not visible to the solution output. Hidden layers are intermediate layers where computation is done. In deep artificial neural networks (i.e., deep learning) there are sometimes hundreds of hidden layers. The network is fed sample inputs and desired outputs

---

91. Apple, Amazon, Alphabet, Microsoft, Facebook, Alibaba, and Tencent began by using "shallow" technology to try to solve a problem. Similar to Francesco Gadaleta, Peng T. Ong at Monk's Hill Ventures says you don't look for deep tech when investing or starting a company; instead, find deep value. Deep tech will emerge if it becomes necessary or valuable.

over and over, and the network learns to fit a function by adjusting the weights in each hidden layer. What each layer learns is ultimately part of the desired output.

The weights in artificial neural networks are updated using so-called back-propagation ("backprop" for short). Geoffrey Hinton's breakthrough was to show that backprop could train a multilayer artificial neural network and that it can—in a general sense—learn useful distributed representations of data by back-propagation of errors in the hidden layers of artificial neural networks. However, the distinction advocated by some literary experts is a distinction that by itself does not deliver the goal promised. It is easy to fall into the trap of abstracting away a problem, believing that we can "simply stack an arbitrary number of hidden layers together" and backprop will solve any problem.[92]

Ultimately, there are a host of complex trade-offs among complexity (i.e., making solutions bigger and/or deeper), parallelism (e.g., data and algorithm), accuracy, generalization, resources (i.e., specialized hardware), parameter and hyperparameter tuning, resource utilization (e.g., memory), and human resources (i.e., personnel). In fact, if you lack comprehension of your problem, infrastructure, resources, teams, customers, deployment environment, or your baseline solution, then stacking more layers will not fix anything. If you have lots of dense data, such as text or images, and need to increase performance, you may have no choice but to use an inscrutable solution with hundreds of layers and millions of parameters. However, bigger and deeper is not always better.

Generally, smaller and simpler is better. The goal of reducing complexity and finding minimum performance thresholds is often preferable to seeking more complexity and maximum performance thresholds. Regardless, solution complexity is not the strongest predictor of solution performance. In fact, a problem and its complexity will better predict solution performance. That is, high performance is better predicted by easier

92. A. Karpathy, "Yes, You Should Understand Backprop," *Medium*, https://medium.com /@karpathy/yes-you-should-understand-backprop-e2f06eab496b.

problems captured with less complex data than it is by solution complexity. The point is, if we do not require exceedingly high performance—or if we lack a complex problem with complex data, lack Big Data, lack lots of memory, specialized personnel, hardware, and time—then we may not want to arbitrarily create a deeper and inherently more complex solution.

In fact, at the heart of deep learning and machine learning you'll often find an optimization problem. The more complex a solution, generally, the harder the optimization problem becomes, which often accompanies a corresponding increase in data in order to generalize. For example, one dreadful characteristic of deep learning is "over-parameterization." As networks become deeper and deeper, the number of parameters is larger than the data used to train the learning algorithm, which creates the potential to overfit on the training data.[93] As any supervised machine learning solution becomes more complex, it performs better against the training data and less so in the real world because generalization is compromised.[94]

Silver-bullet thinking reflects a desire to have answers or to have a solution before we have asked a question or noticed a problem. By rearranging the temporal ordering of problems and solutions, silver-bullet thinking is like building a key without a door.[95] We have a key, perhaps the biggest, most complex, and elegant key devised, but we spend all our time searching for the right doors to open. When push comes to shove, business pursuits require more than awareness of solutions or knowing very crude assessments such as vanity counts of hidden layers or volume of data.

---

93. Jianqing Fan, Cong Ma, and Yiqiao Zhong, "A Selective Overview of Deep Learning," *arXiv*, arXiv:1904.05526, April 2019, http://arxiv.org/abs/1904.05526.

Over-parameterization is a special case of the Curse of Dimensionality (CoD). CoD occurs when the dimension of covariates (i.e., possible predictive or explanatory features) goes up—something called *dimension explosion*—while the number of observations stays the same, which is often associated with a solution's efficiency going down. There are certain techniques to handle the dimension explosion, but most offer no guarantee of working.

94. Besides, the neural network zoo underscores that the power of artificial neural network architectures is generated by the number of architectures, not the depth per se.

95. Ash Maurya, "Don't Start with an MVP," Leanstack, March 29, 2018, blog.leanstack.com/dont-start-with-an-mvp-aa883de5cd18.

Many of us still associate learning with memorizing facts, formulas, and concepts, but leadership requires us to notice problems, not just perform vulgar counting of hidden layers in artificial neural networks.[96] Vulgar counting of vanity metrics provides no guidance into how we ought to define business-use cases, convert business requirements into data requirements, continuously improve solutions in terms of accuracy and fairness, or customize machine learning solutions with problem-specific information.[97]

Ultimately, our awareness of a solution is influenced by our imagination, which is influenced by misinterpreted results in restricted domains, aesthetic claims, and fluency heuristics. However, awareness, aesthetics, and fluency heuristics are not accomplishments. In fact, if the amount of data is all we have to talk about, then we are almost certainly failing. We are also hiding failures of problem solving if all we can point to in our capital expenditures is the number of hidden layers within a solution. Instead of accumulating data—or hidden layers—we should consider finding the right problems to solve and then begin working backward to identify what data supports problem solving and determine how we collect it. To be sure, running toward some arbitrary solution often requires us to run away from the problems we were supposed to solve.

---

96. Aytekin Tank, "Why Learning More Isn't Always Better," *Medium*, January 7, 2019, https://medium.com/swlh/why-learning-more-isnt-always-better-2154f75a71d6.

97. Rama Akkiraju et al. discuss how academic literature on machine learning modeling fails to address how to make machine learning models work for enterprises. Akkiraju and team provide a reinterpretation of the Capability Maturity Model for the machine learning development process. Rama Akkiraju, Vibha Sinha, Anbang Xu, Jalal Mahmud, Pritam Gundecha, Zhe Liu, Xiaotong Liu, and John Schumacher, "Characterizing Machine Learning Process: A Maturity Framework," *arXiv*, November 12, 2018, arXiv:1811.04871, https://arxiv.org/abs/1811.04871.

ALL WE NEED IS MORE TIME

Tomorrow is tomorrow. Future cares have
future cures, and we must mind today.
—Sophocles

The tension between insiders and outsiders is often the tension that exists between the solutions we have today and the solutions someone else wants in the future. In fact, the dynamic between what our solutions are and what our solutions will be is core to artificial intelligence. AI has always been this way. It's an area of research with an eye on the future, maybe two eyes. The AI Effect highlights this horizon-seeking approach, as solutions are seen as stuck in the present and evaluated against whatever they haven't yet done.[1] Business, however, can't keep two eyes on the future.

---

1. Enter "artificial intelligence" into Google Search and you'll find as a related search "artificial intelligence future" (search performed 2/22/20).

For this reason, I am surprised by the frequency at which I am asked to predict where artificial intelligence will be in five or ten years.[2]

We ought to recognize that such a prediction about the future of our solutions—like nearly all predictions about the future—will be wrong. The question of where artificial intelligence will be in five or ten years reflects something odd, which is that artificial intelligence seems to have made it easier to talk about the future of solutions more than our present problems. AI causes many to seek answers from the future rather than the present.[3] We generally desire solutions ahead of their time, but that notion is almost always incoherent. Incoherent because the future is not a source of knowledge that can ever be validated and because we cannot measure progress against ideals of the future, only against where we start solving a problem. To be sure, tension builds if we ruminate too much about the future of our solutions.

It's hard to pinpoint our collective love affair with the future, yet many of us create a strange attitude about the present. Consider that if we anticipate "real" artificial intelligence in the next five or ten years, then we may spend little or no time ruminating on today's problems. Here's the trap of the future solution: if you were told that general intelligence would happen in your lifetime, then your actions would be different because your vision of the future would be different. The way we invest in our personal financial futures, for example, is different when we are twenty than when we are sixty. In a real way, our stance toward the future is our future. To be sure, anyone talking some abstract future with abstract value is most unlikely to actually get there because (1) we are part of our future, which is influenced

---

2. This was a question I was asked by the NSTAC subcommittee in 2018. *See* "Cyber-security Moonshot Initiative Action Plan," in https://www.cisa.gov/sites/default/files/publications/NSTAC_CyberMoonshotReport_508c.pdf.

3. This is not new: ABC had a show titled *Tomorrow* that ran in 1961, talking about the future of AI. More than sixty years later and little has changed because we have not changed. *See* @BrianRommele, Twitter, https://twitter.com/BrianRoemmele/status/1227591424931061763?s=20.

by our actions today, and (2) knowledge requires explanation, and explanation of future technology is always dubious.

Future advantages do not require futuristic solutions, only better solutions in the present. The tendency to imagine that all we need is more time is dangerous, particularly for business because forecasting can lead to becoming promiscuous about problem solving. Forecasting is passive, rather than active. If all we have to do is wait for the future, then we will never be wrong while we wait. The habit of looking forward to the future is hard to escape, but business is not a waiting room.

Our lives are absolutely riddled with many different modes of waiting. We wait for luck to turn in our favor, which it often doesn't. We wait for apologies that we are unlikely to get. We wait for others to notice us or our efforts, but they don't. In *The Monk and the Riddle* Randy Komisar discusses the "deferred life plan," where we wait for a mega-round of funding, that one big customer, or a large exit so that we may find problems we are truly passionate about solving. We wait for things to be perfect even though nothing can be. We wait for perfect solutions, rather than building a good one. We wait for the future even though the present is all that we have. Ultimately, the present provides us with a future opportunity, whereas the future cannot provide an opportunity today.

AI can only be what it is. If we need it to be something else, its deficiencies should be based on the nature of our problem and the gaps in solution performance. This is precisely when organizations have an opportunity to be creative, driven by what a solution might need, to solve a problem. Economist Theodore Levitt wrote about this in a timeless *Harvard Business Review* article titled "Creativity Is Not Enough."[4] He advises focusing on the right problems instead of being creative for creativity's sake. This makes intuitive sense as it relates to AI because a perfect solution does not need to include everything to solve intelligence, only what is necessary

---

4. Theodore Levitt, "Creativity Is Not Enough," *Harvard Business Review*, August 2002, hbr.org/2002/08/creativity-is-not-enough. In simpler terms, customers want to "hire" a solution to do a "job," or, as Levitt put it, "People don't want to buy a quarter-inch drill. They want a quarter-inch hole!"

to solve a problem. If you don't know anything about a problem, then you won't know anything about what a customer might need.

A contributing factor to the confusion around AI is that many books on artificial intelligence focus on the uncertain future of our solutions. Consider that Amazon has more than thirty thousand books covering the topic of artificial intelligence, which in and of itself is quite astounding. There is a huge volume of information that's readily available, but much of it is mired in superficial understanding, facile conventions, and pompous maxims. Many of these books talk about what our solutions need to look like, will look like, and what the collective ought to worry about in the future. We read, and we wait, and we watch as the bad drives out the good. Meanwhile, our problems don't wait to be solved. They continue being problems. How much *longer* do we wait? What are we waiting for? When does perfect become the enemy of good?

Much of the artificial intelligence orthodoxy is the inspection of what our solutions need according to a decades-old goal. We live with false hope and a strange uncertainty. Consider GitHub and the platform's near 31 million developers. Collectively these talented developers have created one hundred or so repositories that match the search "artificial general intelligence." I suspect this ratio of more than 300 to 1 (books *to* repositories) highlights that it is easier to write a book about artificial intelligence—*ahem*—than it is to design intelligence into a program. To be sure, none of these repositories actually contain a program capable of general intelligence. In fact, a major source of simple-mindedness is referring to these solutions by their grandiose pursuit.[5]

The epidemiology of the problem is blurry, but professor of physics at the University of Michigan, M. E. J. Newman, suggests why insiders tend to promote the future.[6] Newman finds a strong first-mover effect in which

---

5. This point was also made by Drew McDermott in his 1976 paper titled "Artificial Intelligence Meets Natural Stupidity."

6. M. E. J. Newman, "The First-Mover Advantage in Scientific Publication," *Europhysics Letters* 86, no. 6 (June 2009), https://iopscience.iop.org/article/10.1209/0295-5075/86/68001.

the first academic papers in a field—regardless of content—will, essentially, receive citations at a rate enormously higher than papers published later. Newman adds that the scientist who wants to become famous is better off—by a wide margin—writing a modest paper in next year's hottest field than an outstanding paper in this year's.[7]

Newman illustrates that the biggest payoffs understandably go to the pioneers. However, like moths to a light, insiders are often attracted to what is seen as the future, so they create variant solutions. But these variant solutions get much less attention. According to the law of diminishing returns, every variant of tomorrow's solution gets less and less attention today. For example, the director of machine learning at Apple, Ian Goodfellow, is famous for inventing Generative Adversarial Networks (GAN) for his PhD thesis at the University of Montreal in 2014.[8] Now there are more than five hundred variants of GANs.[9]

Outsiders must understand that insiders do not require definite end goals because they operate in unconstrained environments whereas business operates in constrained environments. Where does it end? As University of Massachusetts professor, and director at Adobe Research, Sridhar Mahadevan asks, "What is the point of the 501st GAN, 401st regression method, 301st policy gradient method for deep reinforcement learning, 251st clustering algorithm, or the 151st variant of stochastic gradient descent?"[10] Managers have to get things done and should avoid the 501st variant of a solution. The biggest investment advice for business-minded

---

7. Ryan Hill and Carolyn Stein find evidence to support that getting *scooped* will damage the likelihood of publication and the likelihood of appearing in a top journal, as well as the likelihood of fewer citations, though these likelihoods are much more moderate than scientists actually believe.

8. Ian J. Goodfellow et al., "Generative Adversarial Networks," *arXiv*, June 2014, arXiv: 1406.26612014, http://arxiv.org/abs/1406.2661.

9. A. Hindupur, "The GAN Zoo," *Github*, https://github.com/hindupuravinash/the-gan-zoo.

10. Sridhar Mahadevan, "Does Every Paper in Machine Learning Introduce a New Algorithm?" *Quora*, August 31, 2020, www.quora.com/Does-every-paper-in-machine-learning-introduce-a-new-algorithm/answer/Sridhar-Mahadevan-6?ch=10&share=4657945d&srid=1Rf.

leaders is to put your money—so to speak—in what you know. Managers know their business and customers. Focus on the baskets, not the eggs.

I am not advocating that insiders restrict their writings, place a cap on the number of variant solutions, demand some governance over futuristic papers, or demand that application be their sole focus.[11] The friction that would result from these imagined mechanisms would slow down insiders, reduce the exchange of ideas, and ultimately interfere with new solutions and new methodologies. However, outsiders need to interpret futuristic claims for what they are: someone else's orthodoxy. Ultimately, we must comprehend the impact of today's solutions on today's problem and anticipate their impact on the future, not the imagined impact of future solutions satisfying abstract, external goals.

How Fast Are We Going?

Distinguished scientist at SRI International, Charles Perrault, is one of the leaders of the so-called AI Speedometer for a project named the AI Index, which aims to track progress in the field of AI.[12,13] The AI Speedometer is effectively tracking the flow of micro-discoveries, mostly through published research papers on machine learning, while suggesting that these

---

11. In fact, I strongly disagree that basic research should be substituted with applied research. Insiders do not merely seek answers but also seek theory. Theory is important to explain why progress in a domain is better than previous answers. A movement from basic research toward applied technological research would, as Jonathan Zittrain has written, "certainly threaten academia's investigative rigor or bypass them entirely." (*See* "Intellectual Debt: With Great Power Comes Great Ignorance," https://medium.com /berkman-klein-center/from-technical-debt-to-intellectual-debt-in-ai-e05ac56a502c.) I am nearly saying that businesses often only need answers and not theory.

12. The project is backed by the "One Hundred Year Study on Artificial Intelligence," established at Stanford in 2015 to examine the effects of AI on society.

13. Tom Simonite, "Do We Need a Speedometer for Artificial Intelligence?" *Wired*, February 2, 2018, www.wired.com/story/do-we-need-a-speedometer-for-artificial-intelligence/.

small victories are significant enough to result in big achievements. This is unlikely to be true.[14]

Scientist and author Gary Marcus defines micro-discoveries as the steady stream of self-promotion via small victories.[15] He suggests that the media promotes these small advances as though they will soon change the world, but micro-discoveries are about overconsumption—a type of solution gluttony—and resonate because we often cannot resist the allure of progress. The barrage of micro-discoveries leaves us trained not just to expect victories but also to demand victories at a dizzying pace.

On one hand, we must remember that following a breakthrough—like the 2012 AlexNet victory in ImageNet—we can only expect small victories and micro-discoveries. Consider that as performance approaches 100 percent accuracy on a problem, improvements will naturally slow and shrink because we cannot exceed 100 percent accuracy. Remarkably, performance improvements in problems like image recognition, for example, continue to increase roughly linearly.

On the other hand, a recent paper by Kevin Musgrave, a PhD student at Cornell University, shows that when solutions are properly tuned, most of them perform similarly to one another.[16] Musgrave's research effectively states that when properly configured and properly compared, using cross-validation, many solutions perform exactly the same despite "state-of-the-art" claims, because those who make a state-of-the-art

---

14. Hubert Dreyfus called this idea the "first-step fallacy" and claimed that "ever since our first work on computer intelligence we have been inching along a continuum at the end of which is AI so that any improvement in our programs no matter how trivial counts as progress." Hubert L. Dreyfus, "A History of First Step Fallacies," *Minds and Machines* 22, no. 2 (2012): 87–99.

15. Gary Marcus, "Artificial Intelligence Is Stuck. Here's How to Move It Forward," *The New York Times*, July 29, 2017, www.nytimes.com/2017/07/29/opinion/sunday/artificial-intelligence-is-stuck-heres-how-to-move-it-forward.html.

16. Kevin Musgrave, Serge Belongie, and Ser-Nam Lim, "A Metric Learning Reality Check," *arXiv*, March 2020, arXiv:2003.085052020, http://arxiv.org/abs/2003.08505.

claim are not configuring rival solutions properly and thus are making unfair comparisons.[17]

More importantly, while progress against benchmark data sets have been impressive, we must acknowledge that progress in closed domains producing linear, sequential units do not result in the achievement of ambiguous goals like solving intelligence, rendering so-called speedometers largely in vain.[18] In fact, history has shown that attempts to reach general intelligence by expanding narrow intelligence with small victories have made no headway over many decades of research. If there is any doubt, consider that IBM, who created both Deep Blue and Watson, is no closer to solving intelligence than anyone else.

This explains what is difficult to ignore: we are no closer to general intelligence even as insiders get closer and closer to 100 percent accuracy on benchmark data sets using cross-validation. Cross-validation is a way to measure how well a solution generalizes by testing a solution against data withheld in a training set. However, benchmark data sets and cross-validation are "not a viable way to construct a 'general-purpose' solution."[19] This explains how the AI Speedometer can be pegged, yet real artificial intelligence is nowhere to be found.[20]

All of this is to say that insiders use problems, metrics, and testing that serve themselves, which doesn't make insiders evil; it makes them

---

17. Furthermore, determining statistically significant change is difficult because most claims do not perform statistical testing to provide any sort of evidence that differences in model performance is statistically meaningful.

18. ANI, AGI, and ASI do not exist on some strange continuum that would justify monitoring speed because progress is not linear.

19. An algorithm designed to apply beyond a specific task but without a specific, non-trivial characterization of the class(es) to which it should be applied. *See* Christophe Giraud-Carrier and Foster Provost, "Toward a Justification of Meta-learning: Is the No Free Lunch Theorem a Show-stopper?" in Proceedings of the ICML-2005 Workshop on Meta-Learning.

20. In other words, insiders do occasionally measure progress from where they started via contests and measure where they want to go via human thresholds, but accuracy and human thresholds are poor proxies for intelligence, and benchmark data sets are poor proxies for real-world problems.

goal oriented. We must understand, however, that self-serving problems and metrics are not specifications of intelligence. Even when taking on the most favorable interpretation of this work, we must remember that the value of measuring speed when you don't know the destination or how to get there is never worth the amount of time and effort that goes into measuring it. All of which explains why so-called AI Speedometers are useless.

Recall that insiders do not agree on where they are, where they are going, or how good a given solution is for some disputed problem.[21] Consequently, knowing the speed of something is useless if you don't know where you are,[22] where you are going, the distance to travel, and the time that one has to reach a destination. In fact, going fast or faster means next to nothing if you don't know where to go, or how or when to arrive. Moreover, when distance and time are known and fixed, but speed varies due to research hurdles, speed alone will not be enough to accomplish a goal no matter how fast one goes. The AI Speedometer is in effect saying, "We're lost, but we're making good time!"[23]

Instructional frivolity aside, the real paradox is that slowing down may help insiders go farther much faster than going fast. As the Slow-Science Movement Manifesto underscores:

> Science needs time to think. Science needs time to read, and time to
> fail. Science does not always know what it might be at right now. Sci-
> ence develops unsteadily, with jerky moves and unpredictable leaps
> forward—at the same time, however, it creeps about on a very slow

---

21. Drawing a real-world comparison, Peter Sweeney asks rhetorically to imagine a military commander who is unable to understand the enemy, or a CEO who doesn't understand the market. P. Sweeney, "Sharpening the AI problem," *The Explainable Startup*, March 6, 2021, https://www.explainablestartup.com/2019/07/sharpening-the-ai-problem.html.

22. MIT professor and director of MIT's Center for Bits and Atoms Neil Gershenfeld notes how discussions of artificial intelligence have been oddly ahistorical. In his essay in *Possible Minds*, he compared the discussions of AI to a mental health condition. He writes how insiders "could better be described as manic-depressive: depending on how you count, we're now in the fifth boom-and-bust cycle." (J. Brockman, ed., *Possible Minds: Twenty-Five Ways of Looking at AI* (New York: Penguin Books, 2019), 162.)

23. As Yogi Berra said: https://quoteinvestigator.com/2012/07/11/making-good-time/.

time scale, for which there must be room and to which justice must be done. Slow science was pretty much the only science conceivable for hundreds of years; today, we argue, it deserves revival and needs protection. Society should give scientists the time they need, but more importantly, scientists must take their time.[24]

Of course, pundits would have much less to write about, and Twitter would be much less exciting if things did slow down.[25] But it does not change the fact that the accumulation of scientific knowledge doesn't require a speedometer. Thomas Kuhn perfectly describes the manner in which scientific knowledge is developed. Knowledge is not based on accumulation or even effort. It is not entirely reliant on one person or spawned on a specific day.[26] Scientific revolutions occur by a process that involves the intervention of various people and events, all of which, if analyzed in isolation, might appear arbitrary or even insignificant.

Said differently, discovery does not occur in straight lines, and attempting to identify the speed or the exact date when a paradigm shift occurs is futile. Whether discoveries are large or small or progress is fast or slow (or any enumeration of these qualities), business thinking ought to be more terrestrial. Consider that exposure to fast-food signs can make you impatient.[27] Working in a dimly lit room results in more cheating

------

24. Slow-science.org, http://slow-science.org.

25. J. Horgan, "The 'Slow Science' Movement Must Be Crushed!" *Scientific American*, July 29, https://blogs.scientificamerican.com/cross-check/the-slow-science-movement-must -be-crushed/.

26. Precisely, Kuhn wrote that it is "difficult to see scientific development as a process of accretion. The same historical research that displays the difficulties in isolating individual inventions and discoveries gives ground for profound doubts about the cumulative process through which these individual contributions to science were thought to have been compounded." Thomas Kuhn, *The Structure of Scientific Revolutions* (Chicago: University of Chicago Press, 1962), 3.

27. Chen-Bo Zhong and Sanford E. DeVoe, "You Are How You Eat: Fast Food and Impatience," *Psychological Science* 21, no. 5 (2010): 619–622.

than similar activities in a lighted one.[28] And serving food on larger plates will make us eat more.[29] The environments we create and keep can and often do influence our thinking and what's more our behaviors. Ruminating endlessly about the future of artificial intelligence can distort our thinking, warp our focus, and corrupt our organizational culture. Ultimately, we cannot conflate our goals with the goals of others. We need to measure progress against where we start, not against where we want to go, and especially not where someone else wants to go. Ultimately, micro-discoveries become a clever way to distort our senses by distorting our perception of time and progress.

Singularity

Consider our general fascination with futuristic visions, championed by futurists like Ray Kurzweil. Kurzweil's 2005 book *The Singularity Is Near* was a *New York Times* bestseller and a number-one book on Amazon in both science and philosophy. To be sure, however, the book is far more philosophy than science.

In general, mathematicians use singularities to describe a problematic point in their research. There are all sorts of singularities, and they're most commonly used to describe a point that reaches an infinite value. American inventor and Google engineer Ray Kurzweil reserves the meaning of the term for something else. For Kurzweil the singularity is a rapid increase in intelligence. Kurzweil writes that "[The] Singularity will allow us to transcend these limitations of our biological bodies and brains . . . There will be no distinction, post-Singularity, between human and machine." Kurzweil refers to this phenomenon specifically as a technological singularity.

---

28. Maria Konnikova, "Inside the Cheater's Mind," *The New Yorker*, October 31, 2013, www .newyorker.com/tech/annals-of-technology/inside-the-cheaters-mind.

29. Alex Swerdloff, "Serving Food on Larger Plates Will Make You Eat More," *Vice*, September 15, 2015, munchies.vice.com/en_us/article/xy74nd/serving-food-on-larger-plates -will-make-you-eat-more.

Kurzweil predicts that by 2029 computers will have human-level general intelligence and a singularity will occur by 2045.[30] Kurzweil asserts that the singularity is emergent, the result of multiplying intelligence a billion-fold. However, these predictions are very difficult to evaluate, and they ask "emergence" to do most of the heavy lifting.

Kurzweil writes that the singularity is near because we are at the "knee of the curve" of some important trend that is growing exponentially. Kurzweil naively relies on exponential growth and says that the "knee of the curve" is a point in time when "events erupt at an increasingly furious pace."[31] However, there is no such thing as a "knee" to an exponential function and any knee, if it were to exist, would be apparent only after, not before, such growth.[32] If we are in front of the knee, then how are we supposed to evaluate a trend before growth? Ultimately, we should be suspicious of anybody whose claims ride the coattails of exponential growth.

Kurzweil compares his technological singularity with that of a mathematical or astrophysical singularity. Although his ideas of a singularity are not actually infinite, he says it looks that way from any limited perspective.[33] In other words, the technological singularity is not a true exponential, which should come as no surprise because exponentials in the physical world are not often true exponentials. These functions are often S-curves. S-curves are flatter at the beginning and end and steeper in the middle. But they may still be neither. Recall that Chollet says that no "intelligence explosion" will occur and that a coevolution of digital and natural intelligence has already been happening and will continue to happen at roughly a *linear pace*—neither exponential nor an S-curve—as intelligence moves

30. Christianna Reedy, "Kurzweil Claims That the Singularity Will Happen by 2045," Futurism, October 16, 2017, futurism.com/kurzweil-claims-that-the-singularity-will-happen-by-2045.

31. Ray Kurzweil, *The Age of Spiritual Machines* (New York: Penguin Books, 1999), 24, Kindle edition.

32. Wayne Yamamoto, "We Don't Understand Exponentials," *Kazabyte*, December 19, 2011, www.kazabyte.com/2011/12/we-dont-understand-exponential-functions.html.

33. Ray Kurzweil, *The Singularity Is Near* (New York: Viking Books, 2005), 23.

to an increasingly digital substrate. Predicting the future with linear trends suggests that if Chollet is correct, tomorrow's technology will look similar to today's technology. The future may not turn out to be as radically different and distinct as we think.

To be sure, infinite values from exponential functions exist in mathematics, but they do not exist in the physical world. In 1925, German mathematician David Hilbert, one of the most influential mathematicians of the late-nineteenth and early-twentieth centuries, wrote about the unphysical nature of infinity. He wrote that infinity is needed to complete mathematics, but it doesn't appear anywhere in the physical universe. Kurzweil seems to treat infinity as just another big number, but the very nature of infinity is quite unlike any other number. Infinity can never be realized. No matter how long we wait or what we do, it is always beyond access.[34] In other words, "anyone who believes that exponential growth can go on forever in a finite world is either a madman" or futurist.[35]

More importantly, tracking trends and the unquestioning belief that we can predict the future from them is not a fundamental understanding of what mechanisms and processes contribute to trends. That is, trends are not causes, they are results. Results break down and cannot be replicated if we don't understand the causes. The idea of a technological singularity is romantic and fuels science fiction, but it's used to grab attention and ultimately create value for the prognosticator and their culture. The technological singularity is philosophy. It's not good science or good business (or business at all), because futurists are spiritual leaders, not business leaders.

---

34. These phenomena should not be confused with general trends of commoditization, defined as the process by which goods that have economic value and are distinguishable in terms of attributes [uniqueness or brand] end up becoming simple commodities in the eyes of the market or consumers, as can be found in Moore's Law.

35. Attributed to Kenneth Boulding in U.S. Congress, *Energy Reorganization Act of 1973: Hearings, Ninety-third Congress, First Session, on H.R. 11510* (U.S. Government Printing Office, 1973), 248. Author added a twist by including *futurists*. The original quote is "Anyone who believes that exponential growth can go on forever in a finite world is either a madman or an economist."

Futurists do not have to live with the consequences of their prognostication, but business leaders do.

A more cynical view is that futuristic predictions are less prophetic than they are self-serving. Such predictions reside within the lives of those who make them. Consider that when we imagine achieving a future goal, we imagine a future that includes ourselves. After all, we can't continue to pursue goals when we are no longer here. Our sense of purpose largely comes from making goals that we can influence or manipulate. Thus, self-serving optimism leads to hedonic prediction that makes the futurist feel good. These types of predictions often reflect a desire of the futurist to live in the world they seek, not because they can influence or understand that world, but because it makes them feel less helpless.

Kurzweil isn't alone in this sense. Predictions are often thirty to forty years forward.[36] The first wide-scale popularizer of the technological singularity was science fiction author Vernor Vinge, who predicted in 1993 that the singularity would occur in 2030, with a time horizon of thirty-seven years and a maturation age of eighty-seven years old. In 2011, Kurzweil predicted the singularity in 2045, with a time horizon of thirty-four years and a maturation age of ninety-seven years old.[37] In 2016, Masayoshi Son, chairman and CEO of SoftBank, predicted the singularity in 2047, with a time horizon of thirty-one years and a maturation age of eighty-nine years old.

The result when balancing forecaster herding, self-serving optimism, and hedonic prediction are claims that are almost always in the thirty- to forty-year futurist sweet spot.[38] The boldest prediction, however, is

36. A review of ninety-five predictions from 1950 have routinely produced results predicting AGI in the next twenty years: *See* Stuart Armstrong and Kaj Sotala, "How We're Predicting AI—or Failing to," in J. Romportl, E. Zackova, and J. Kelemen, eds., *Beyond Artificial Intelligence* (Switzerland: Springer International, 2015).

37. Keep in mind that Kurzweil wants to live forever and spends about $1 million per year on extending his life. *See* https://www.businessinsider.com/google-futurist-ray-kurzweil-live-forever-2015-4.

38. Stuart Armstrong, Kaj Sotala, and Seán S. Ó hÉigeartaigh, "The Errors, Insights and Lessons of Famous AI Predictions—and What They Mean for the Future," *Journal of*

that in the future someone will still be trying to predict the future, only less effectively. As former option trader, author, and statistician Nassim Nicholas Taleb notes—building on Benoit Mandelbrot's Lindy Effect—expectations for the future can often be expressed by power laws, so the longer we wait for something, the longer we can expect to wait.[39] In other words, in thirty to forty years we will still be safely thirty to forty years away from the singularity.

These predictions project something tangible on an otherwise empty horizon. They communicate to the reader or listener that we are close enough to touch something dramatic but not so close to be worried. In other words, while we wait for the future, making predictions about it creates new possibilities that keep us safe from danger or action, even as our lack of action is the most pernicious part of the equation. Said differently, futuristic predictions are often just business code for the fact that we don't have to do anything today. Nevertheless, one of the largest retrospective reviews of technology forecasts over the last fifty years found that forecasts with time horizons beyond ten years were rarely better than coin flips.[40]

Futurists seek to live in the future, but we should always be leery of following them. Do not forget that all futurists have to make oversized claims: it's in the job description (if one had to apply for such a job rather than merely adding it to their Twitter profile). If a futurist makes claims that are moderate and reasonable, then their observations will seem obvious in the present, and over some arbitrary amount of time it will be seen as absurdly conservative should they be correct. Consequently, claims are

Experimental & Theoretical Artificial Intelligence 26, no. 3 (2014): 317–42, doi:10.1080/09 52813X.2014.895105.

39. Kyle VanHemert, "Why Are Old Technologies So Hard To Kill? Nassim Taleb Has a Theory," *Fast Company*, July 9, 2018, www.fastcompany.com/1671964/why-are-old -technologies-so-hard-to-kill-nassim-taleb-has-a-theory.

40. S. Charbonneau, S. Fye, J. Hay, and C. Mullins, "A Retrospective Analysis of Technology Forecasting," American Institute of Aeronautics and Astronautics, August 13, 2012, https://doi.org/10.2514/6.2013-5519.

deliberately extreme and unreasonable; otherwise, the futurist is not much of a futurist.

We need to realize that such claims reflect poor judgment even when they turn out to be true. If Kurzweil (or any futurist) proves to be correct, it will not reflect good judgment, because to predict such a rare event with confidence, one would axiomatically have to ignore available information and rely on intuition, or worse, vanity, to make such an extreme and rare prediction. In other words, futurists know nothing more than the rest of us about the future; they simply possess poorer judgment.[41]

The same type of idealism is encapsulated in notions of BHAGs,[42] blue skies, and moonshots that have undefined problems, infinite timelines, and exaggerated expectations. Idealism is the point of view that solutions must be from the future, magical, or verging on the impossible. They do not. Businesses do not have to solve logical paradoxes such as discovering the possible by learning the impossible.[43] Our work does not need to be impossible, because when it is, we justify our failure to the magnitude of the task rather than admitting our own role in creating the wrong goal and the wrong culture.

For example, in 2017, Gary Marcus argued that AGI needs a moonshot. He said, "Let's have an international consortium kind of like we had for the large hadron collider. That's seven billion dollars. What if you had $7 billion that was carefully orchestrated towards a common goal?"[44] Demis Hassabis describes his company as a kind of "Apollo program effort for

---

41. J. Denrell and C. Fang, "Predicting the Next Big Thing: Success as a Signal of Poor Judgment," *Management Science* 56, no. 10 (2010): 1653–67.

42. Stands for Big Hairy Audacious Goal, an idea conceptualized in the book *Built to Last: Successful Habits of Visionary Companies*, by James Collins and Jerry Porras.

43. British science fiction writer Arthur C. Clarke formulated three adages that are known as Clarke's three laws. They are part of his ideas in his extensive writings about the future. Clarke's second law states, "The only way of discovering the limits of the possible is to venture a little way past them into the impossible." *See* Arthur C. Clarke, *Profiles of the Future: An Inquiry into the Limits of the Possible* (New York: Harper & Row, 1973).

44. A. George, "Discussing the limits of Artificial Intelligence," TechCrunch, April 1, 2017, https://techcrunch.com/2017/04/01/discussing-the-limits-of-artificial-intelligence/.

artificial intelligence." Hassabis's vision of AGI is an easy three-step program. First, understand human intelligence. Then re-create it artificially. Lastly, solve everything else.[45]

What AGI lacks, however, is not more money or fewer steps, but real consensus on the problem.[46] Martin Ford's book *Architects of Intelligence: The Truth About AI from the People Building It*,[47] and John Brockman's book *Possible Minds: Twenty-Five Ways of Looking at AI*,[48] both highlight the lack of consensus. As staff writer at Vox Kelsey Piper observed, "Almost all [insiders] perceive something momentous on the horizon. But they differ in trying to describe what about it is momentous—and they disagree profoundly on whether it should give us pause."[49, 50]

Another reason why BHAGs, blue skies, and moonshots fail us isn't only because of their strange optimism, the lack of consensus, or their uncommonly aggressive makeup, but also because they ask us to answer what we would do if there were no limits to what we could do. However, this attempt to re-create the unconstrained environment of insiders fails us in the real world. In the real world, there are always limits, and when limits are significantly relaxed, problem solving is often complicated. That

---

45. https://www.youtube.com/watch?v=Ia3PywENxU8.

46. Very thoughtful article by P. Sweeney, "Sharpening the AI Problem," March 6, 2021, https://www.explainablestartup.com/2019/07/sharpening-the-ai-problem.html.

47. M. Ford, *Architects of Intelligence: The Truth About AI from the People Building It* (Birmingham, UK: Packt Publishing, 2018).

48. J. Brockman, ed., *Possible Minds: Twenty-Five Ways of Looking at AI* (New York: Penguin Books, 2019).

49. K. Piper, "How Will AI Change Our Lives? Experts Can't Agree—and That Could Be a Problem," March 2, 2019, https://www.vox.com/future-perfect/2019/3/2/18244299/possible-minds-architects-intelligence-ai-experts.

50. IBM CEO Virginia Rometty called Watson a moonshot. Of course, such a claim is overstated as Watson was mainly on display in the comfortable confines of the company's "immersion room." Watson never failed to impress in the immersion room but performed poorly outside of it. It is strange to call your effort a moonshot when you create a demo that can't even leave the lab. E. Strickland, "How IBM Watson Overpromised and Underdelivered on AI Health Care," *IEEE Spectrum*, April 2, 2019, https://spectrum.ieee.org/biomedical/diagnostics/how-ibm-watson-overpromised-and-underdelivered-on-ai-health-care.

is, too much time and too much optimism come with too many options, whereas constraints force decision making, which forces consensus.[51]

In other words, big-budget projects become everything.[52] Big projects come with big investments that are put in place to avoid mediocrity, which paradoxically is the very thing that causes mediocrity. After all, with big budgets, big resources, infinite timelines, impossible goals, and no requirement for customer development there is no reason to say no to anything. But the cost of success at all cost, is it costs too much. People get burned out when there is no horizon to be found. It's really hard to work year after year with no real goals and no substantive problem to solve. Ultimately, we need guards against this type of idealism with SMART goals.[53] SMART goals are Specific, Measurable, Achievable, Relevant, and Time-bounded.

To be sure, we do not need future solutions for advantages or to discover the impossible to learn of the possible. Ultimately, this kind of idealism can't get anything done just as pragmatism can't imagine a bold

---

51. This is similar to the second-system problem in computer science. When you create a blank slate, your efforts will typically be too ambitious. Try to put as much on your slate as you can.

52. We can find some of the same type of idealism encapsulated in planning like The Offset Strategy, by the US Department of Defense (DoD) and Intelligence Community. The belief is that the next "offset" will bring us one step closer to world dominance, even though the perfect solution has eluded us so far. Steve Blank calls this type of thinking "The Offset Dilemma," and he reminds us that future solutions do not lead to advantages, but rather action today leads to advantages. Similarly, the Defense Advanced Research Projects Agency (DARPA)—the research and development agency for the DoD—believes that the next "wave" of solutions will provide us the advantages we need for national security and global advantages. In 2018, DARPA announced a $2-billion commitment to new and existing programs that will develop the next "wave" of AI technology. The so-called third wave of technological advance is one of contextual adaptation and reasoning capabilities, or, said differently, AGI. DARPA operated between 1983 and 1993 with the goal of AGI and spent nearly $1 billion of taxpayer money before the goal of AGI disappeared entirely from Strategic Computing Program by 1990. AI is ahistorical. (Alex Roland and Philip Shiman, *Strategic Computing: DARPA and the Quest for Machine Intelligence, 1983–1993* (Cambridge, MA: MIT Press, 2002); Severo M. Ornstein, Brian C. Smith, and Lucy A. Suchman, "Strategic Computing," *Bulletin of the Atomic Scientists* 40, no. 10 (1984): 11–15, doi:10.10 80/00963402.1984.11459292.)

53. G. T. Doran, "There's a S.M.A.R.T. Way to Write Management's Goals and Objectives," *Management Review* 70, no. 11 (1981): 35–36.

future. Author Simon Sinek says there is a delicate blend of both. The delicate blend is realism. Ultimately hopeful but not idealistic, skeptical but not pessimistic.

The Next 10,000 Start-Ups

Consider Kevin Kelly, author of *The Inevitable: Understanding the 12 Technological Forces That Will Shape Our Future*, who predicts that "AI is already here, it's real, and it's quickening." Similar to Sundar Pichai, Kelly anticipates that a technological force will include everything becoming smarter using cheap, cloud-based, artificial intelligence.[54]

Kelly, however, is an odd duck. He isn't talking about the imagined impact of future technology but rather anticipating the impact of today's technology in the future. Kelly asserts that the formula for the next ten thousand start-ups is to take something that already exists and add AI to it.[55] He adds that we're going to repeat that by one million times, and it's going to be huge. If Kelly's assertion is true, then AI is not hyped at all. Yet his view is still much more realistic than Kurzweil's. The fact that a solution is not from the future and is not so general as to replace the launch of the next million start-ups does not mean that such a solution lacks economic value.

Despite its claim to take something that already exists and add AI to it, Kelly's formula is more about problems than solutions. Kelly, who was a futurist adviser for the movie *Minority Report*, does not advocate for one killer future company. His claim isn't that one future challenger will dominate—which might suggest general intelligence—but rather that

---

54. Kelly's proclamation suggests that companies using machine learning can create long-term sustainable value. This value creation is distinct compared with organizations using traditional software development, where learning aspects of a problem are more vicarious. The ability to learn is an advantage.

55. Frank Kalman, "Kevin Kelly: The Next 10,000 Startups Will Deal With AI," *CLO Media*, July 30, 2018, www.chieflearningofficer.com/2017/02/14/kevin-kelly/. Despite Kelly's lack of specificity on what exactly constitutes "AI," his words suggest machine learning and in many instances deep learning.

the next ten thousand start-ups, "repeated one million times," will operate across every sector and business vertical. This point underscores an
important relationship explored earlier between solution performance on
one problem and solution performance on all problems. Kelly concludes
correctly that high performance on every problem by one company will be
relatively poor. That is, a general-purpose problem solver is as improbable
as a general-purpose company.

Moreover, Kelly's assertion thwarts the notion that future technology
is required to be future proof. His assertion also thwarts the superiority of
"ideas," and that all good ideas are taken, and that businesses need to move
first to win, because he is talking about replacing incumbents by taking
something that already exists and adding AI to it. It's not that all good ideas
are taken, rather most ideas have been executed imperfectly.

To be sure, success is not built upon the future. It's built by executing
better than others in the present. We may ask why challengers would seek
to replace incumbents using AI. Well, it ensures that a business problem
exists, because a paying customer already exists. These customers are paying because they have a problem they cannot or will not solve themselves.
If we're going to perfect the wheel, it is best to ensure that the wheel has
market acceptance in the first place.

Much of innovation involves "making obsolete that which has
been poorly done."[56] Consider Uber, who merely took an existing idea,
transportation—specifically taxi service—and transformed, or "disrupted,"
it to fit today's modern world, just as taxis did one hundred years earlier.
Previously defined problems are just as valuable as freshly defined problems: perhaps more so. Why not pick good, old ideas already validated by
an incumbent that have been executed imperfectly? Consider that a generic
problem is many times more difficult to solve than a well-specified problem, and that old problems are well specified. Before you devote resources

---

56. Jay Abraham, *The Sticking Point Solution: 9 Ways to Move Your Business from Stagna-
tion to Stunning Growth in Tough Economic Times* (Philadelphia, PA: The Vanguard Press,
2010), 37.

to build a solution, you must first find evidence of a problem, specifically a problem that you can monetize.

What is most interesting about Kelly's prognostication is that it measures progress by starting at the point when someone else stopped working rather than by finding an abstract, external goal for a solution. The type of mimicry suggested is probably not the romanticized future many imagine when thinking of the impact of future technology and certainly not the imagined future of AI. However, when we ignore the empty promises of future solutions, we may find that dull prediction is more productive than blind techno-optimism.

For example: Is there anything duller that insurance? Yet, the value of insurance premiums underwritten through machine learning is estimated to be $1.3 billion, and by 2024, that number is projected to be $20 billion. Quantemplate is a London-based machine learning start-up that has raised $25.7 million of total funding to help insurers process data.[57] Ethos, a data-driven life insurance issuer, has recently raised $200 million in their Series D.[58] Clearcover, a platform that uses AI to sell auto insurance, also raised $200 million in their Series D.[59] Flyreel, an AI product for underwriters, raised nearly $10 million in a recent Series A. And, in 2020, Lemonade—a full-stack insurance company powered by AI and behavioral economics—with $480 million of total funding, was rewarded by investors when it IPO'd with a $3.8 billion valuation.

Kelly's formula emphasizes the importance—and ultimately the value—of well-defined problems. In fact, Kelly's assertion to "take something that already exists and add AI to it" highlights that many well-defined problems can be found as by-products of old solutions—a type of "innovator's gift"—whereas business problems are not found in the future of artificial

57. https://www.crunchbase.com/organization/quantemplate/company_financials.

58. https://www.crunchbase.com/organization/ethos-life/company_financials.

59. https://www.crunchbase.com/organization/clearcover/company_financials.

intelligence.[60] This perspective may explain why futurism repeats the same things over and over again.[61] After all, who are "disruptors" other than entrepreneurs that break other entrepreneurs' solutions to old problems?

In fact, if you want to be irrelevant, you should find a problem that no one has had in the past. The idea of the past informing the present suggests that "red oceans" are every bit as commercially viable as "blue oceans," largely because the new is replaced by newer, bluer oceans, whereas old problems have survived random events occurring in bloody, robust red oceans.[62] Although creating duplicate companies and solving old problems may seem to lack imagination, Kelly's assertion highlights an important point, which is that old problems solved better in the present are more important than better, abstract problems solved in the future. You may not be able to predict the next big thing, but you can predict that existing problems solved better today will at least be entertained by existing customers.

Some of us want to believe in the perfect solution so much that we look to the only place where it can exist: the future. However, we risk falling into the same trap in the present: we seek perfect solutions instead of perfect problems to solve. The etymology of the "better mousetrap" is derived from American essayist and philosopher Ralph Waldo Emerson, who wrote, "if a man has good corn or wood, or boards, or pigs, to sell, or can make better

---

60. Ash Maurya, "Don't Start with an MVP," *Love the Problem*, March 29, 2018, blog .leanstack.com/dont-start-with-an-mvp-aa883de5cd18.

61. Malcolm Gladwell discusses the real genius of Steve Jobs is not as a visionary and large-scale inventor but more as a "tweaker." *See* Malcolm Gladwell, "The Real Genius of Steve Jobs," *The New Yorker*, November 14, 2011, www.newyorker.com/magazine/2011/11 /14/the-tweaker.

62. Kim and Mauborgne argue that blue oceans are more viable for business than so-called red oceans. Red oceans denote industries in existence today. Blue oceans denote industries not in existence today. W. Chan Kim and Renée Mauborgne, *Blue Ocean Strategy: How to Create Uncontested Market Space and Make the Competition Irrelevant* (Boston: Harvard Business Review Press, 2010). Kai-Fu Lee makes a similar point with respect to red oceans in his book *AI Superpowers*. Lee asserts (more or less) that quantity is more valuable than quality and that the US entrepreneurs generally seek pure innovation, which is especially true for AI ventures. Lee characterizes Chinese entrepreneurs to be generally focused on the market. Kai-Fu Lee, *AI Superpowers: China, Silicon Valley, and the New World Order* (Boston: Houghton Mifflin Harcourt, 2018).

chairs or knives, crucibles or church organs, than anybody else, you will find a broad hard-beaten road to his house, though it be in the woods." A shortening of this: build a better mousetrap, and the world will beat a path to your door.

For business, this idiom describes more of a curse than a blessing. Even when a better solution exists in the abstract, like "artificial intelligence," it does not mean it will sell when disconnected from a market-relevant problem. We should seek better solutions that are better in a specific sense. A solution is not better until it has been positively evaluated against a real problem—the problem is what makes the solution relevant. Just because we can build a solution does not mean we should, because it does not mean anyone will care.[63]

All of us want to believe in the superiority of a solution and hope a celebration will follow, but this is often not the case. Private technology research firm CB Insights found in a postmortem of failed start-ups that tackling problems that didn't serve a market need was the main reason (42 percent) why they failed. The postmortem highlights that problem guessing using solution-focused strategies is not as effective as problem solving. Treehouse Logic applied the concept more broadly in their post-mortem. Treehouse Logic had a great solution, great data on shopping behavior, a great reputation as a thought leader, great expertise, and great advisers. What they didn't have was a solution that solved a real problem. Better mousetraps are not always hard to build—though a perfect mousetrap is elusive—but they're a lot harder to sell when they ignore market-relevant problems.

Ultimately, the future of our solutions may be the singularity. But not in 2021 or, for that matter, the foreseeable future. In the meantime, we still need to solve the right problems.

---

63. Think of the dot-com bubble.

SOLUTION ARGUING:
A LESSON IN CULTURE,
CONFLICT, AND SPILLOVERS

Nobody made a greater mistake than he who
did nothing because he could do only a little.
—Edmund Burke

On January 2, 2018, Gary Marcus published a critique titled, "Deep Learning: A Critical Appraisal."[1] His central question was whether deep learning could solve general intelligence. It took

---

1. Gary Marcus, "Deep Learning: A Critical Appraisal," *arXiv*, January 2018, arXiv:1801 .006312018, http://arxiv.org/abs/1801.00631.

a mere seventy-two hours for insiders to respond and ignite the first AI Twitter debate of 2018.[2]

This critique and very public debate is highlighted for three reasons:

1) To reveal that insiders are not one group, but rather various groups with their own ideas of what problems to solve and how to solve them
2) To demonstrate how conflict within regimes creates and reinforces internal cultures
3) To explore whether the question "Can deep learning solve general intelligence?" is even important to business

This debate underscores that at least some of the insider conflict is the result of the unofficial nomination of deep learning as the singular methodology for artificial intelligence. In fact, if you used "AI" to describe deep learning, then you have certainly upset some insiders, while (perhaps) making others quite happy. The nomination is unofficial in the sense that the field is fragmented and has not decided that one solution will represent the discipline. The reason is simple. Scientific work is more detailed and requires more explanation than some peevish pundit needs to call a solution AI. Absent of solving intelligence, insiders have more work to do to explain their work.

The unofficial nomination of deep learning as artificial intelligence not only points to its utility but also highlights the conflict that arises from the lack of agreement on the viability of deep learning for ostentatious goals like general intelligence. In this sense, the unofficial nomination has harmed the connectionist movement because it shifts the debate about

---

2. Synced, "Gary Marcus's Deep Learning Critique Triggers Backlash," *SyncedReview* in *Medium*, January 12, 2018, medium.com/syncedreview/gary-marcuss-deep-learning -critique-triggers-backlash-62c137a47836.

what deep learning can do to whether the solution has met the larger, more complex goals of artificial intelligence.

Yet the "is it *real* enough" debate has historically defined the field. This type of "solution arguing" over artificial intelligence represents an important part of the culture and values of insiders, who argue about the name of solutions and ultimately whether a solution deserves a better name or to have a name at all. Certainly, solution arguing is not helpful for problem solving because our solutions do not require names or external goals. As explored in the previous chapter, what is currently successful has to be good enough because we cannot solve a problem with technology we don't have. Yet, for insiders good enough is rarely enough.

Marcus's appraisal of deep learning includes aspersions that are not limited to:

1) Deep learning thus far is data hungry.
2) Deep learning is shallow (shallow in the sense of the program's understanding, not in terms of its architecture).
3) Deep learning possesses limited capacity for so-called transfer learning.
4) Deep learning is not sufficiently transparent.
5) Deep learning thus far cannot inherently distinguish causation from correlation.
6) Deep learning presumes a largely stable world in ways that can be problematic.
7) Deep learning works well as a statistical approximation, but its answers often cannot be trusted. Ultimately, deep learning lacks true understanding.

Marcus's critiques touch on the technical boundaries of deep learning, which we need to be aware of in order to understand the limitations of some subset of solutions, which are not, and may never be, perfect. After all, a solution is only a reliable solution if one knows where it might break down. In this sense, Marcus's arguments are important, and, in all fairness, his aspersions are completely justified as they enable honest assessments of technology. To be sure, deep learning is greedy, brittle, opaque, and shallow. Deep learning is neither magic nor a solution to everything.

Moreover, for the achievement of audacious goals like solving intelligence, insiders must rely on the identification of puzzles. Puzzles will be identified by various experts, including Marcus—ultimately across various groups of insiders—as a basis to develop reasonable expectations for ambiguous and ambitious times ahead. Therefore, comprehension of technical boundaries like those outlined by Marcus requires honest assessments of technology to include anomalies and puzzling results rather than heavy-handed assessments.

Of course, Marcus's critique is general enough to be relevant criticism on much of machine learning and specifically the supervised machine learning paradigm. In fact, his critique may loosely be interpreted as a critique on supervised machine learning (the dominant learning paradigm of machine learning) but directed at deep learning due to deep learning's success and its unofficial nomination as the singular solution for artificial intelligence. In this sense, deep learning has become a victim of its own success.

Marcus asserts unequivocally that deep learning is not perfect. But, I ask, what is perfect?

Like most insiders, Marcus seeks to satisfy the external goals of AI and strives for the perfect solution. However, the desire for the perfect solution is itself an imperfection unless you're an insider, futurist, or madman, because there are no perfect solutions. The Perfect Solution Fallacy is a false dichotomy that incorrectly assumes that such a solution exists. The fallacy is an example of black-and-white thinking, in which someone fails to see the complex interactions between problems and solutions and as a result reduces all solutions down to extreme pairs of perfect and dogshit. This ultimately subverts economic thinking because a perfect solution is only notional and practically unachievable.

In the real world, a "perfect" solution that certainly does not exist is far less valuable than an imperfect, achievable solution we can have today. Perfection is not a value proposition that you should ever seek because it ignores what you need and what customers want. There are many problems (aside from wicked problems, explored more in part two of the book) worth solving adequately, at least for a while.

Consider also that even if we had perfect solutions, they would still create problems. Case in point: when the late English theoretical physicist and cosmologist Stephen Hawking said AI is the potential end of humanity, what he was saying is that perfect still has an assortment of complicated by-products. In other words, even quasi-insiders who entertain perfection still believe that perfection would do harm. Fundamentally, all solutions are a mixed bag of trade-offs. This is especially true given the probabilistic nature of machine learning and statistical learning. That being said, although machine learning isn't a solution to everything, or even a perfect solution for something, amazingly it is a good solution for a lot and shouldn't be overlooked.

To be sure, all solutions are a mixed bag of trade-offs. Consider the paradox of automation, which is a specific class of errors people tend to make in highly automated environments. In these environments, events are handled by automated aids and the human is largely present to monitor things. The paradox is that a well-designed system may hide human weaknesses sufficiently well, which is the intent. However, over time, human weaknesses become dependencies. In mission critical systems these dependencies also become risk factors. That is, if an automated system has an error, it will multiply that error until it's fixed, but humans may not be in a position to fix, or even identify, the malfunction. Consequently, blaming humans for errors in highly automated environments is usually a mistake. Automation creates blind spots and humans may not be in a position when systems fail (and systems will fail) to take over after having lost some keen ability. Human-centric design is obligatory.

Marcus's argument of "not good enough" is according to his own values, which align with his own goals. An argument I frequently hear is,

"Well, it can't do X!," where "it" is some arbitrary solution, and X is often some cognitively plausible feature that is absent.[3] The argument, however, is fallacious because it implies that if a solution can't do X, then it is not useful for performing Y. The same argument ignores that X can't often do Z, and ignores the advantages of Y in the first place.

More importantly, the "it can't do X" argument rarely makes sense, particularly when it ignores a problem or is independent of it. Solutions should not be evaluated according to the standards of others, who often have very distinct and oversized goals, but against a problem. Comparing a solution with intelligence misses the entire point of problem solving. This slip is often associated with naming a solution AI. For example, if we espouse that something is or is not AI "because of X," then we are perpetuating the wrong culture, because everything pertaining to X relates to the insider culture of solution naming and arguing about those names.

Solution arguing occurs when the debate is about the precise definition of a solution, measured against the external goals of that solution. Such goals represent the aspirations of someone else. Ultimately, talking about how much "intelligence" a solution has, or should have, is a trap. Solution arguing has nothing to do with business and certainly nothing to do with problem solving. Nor is it a value proposition that will resonate with most humans. Solution arguing is not the beginning of a journey. It is the end. Solution arguing is where problems die.

Specifically, we must avoid solution arguing when it is sourced from someone else's goals and values. For business, the goal is agency. In social science, agency is the capacity of individuals to act independently and to make their own free choices. To that end, the benefit of owning the name of your solution is agency. Agency shifts power away from scholarly goals,

---

3. Rodney Brooks made a similar argument: *See* Rodney A. Brooks, "Elephants Don't Play Chess," *Robotics and Autonomous Systems* 6, no. 1–2 (1990): 3–15, https://doi.org /10.1016/S0921-8890(05)80025-9.

suitcase terms, insiders, disingenuous celebrities, literary experts, and shoddy media coverage, and gives power back to businesses.

This power comes with responsibility. Agency is a collective model, and you can choose to invest in it or not. In other words, agency comes at a cost. You should only name your solution (if you even name your solution) *after* you solve a real problem. Furthermore, if you choose to name your solution "AI," you should plan to defend the term, because it will likely mean something that you hadn't intended to someone else. By taking control of our solutions, and the names we give them, we also take accountability for solutions, their shortcomings, and the people they impact.

Trust yourself. Trust your problems. Don't trust your solutions.

Good Old-Fashioned Artificial Intelligence

Marcus's agenda is to increase machine capacity to "understand" so that we may more easily trust machine outputs. This type of progress will be important if humanity seeks to impact a wider range of problems, especially those for which the cost of error is high. Ultimately, Marcus promotes a combined paradigm and the unification of different styles (i.e., macro-approaches) of AI to leverage (at least) symbolic representation and sub-symbolic learning (i.e., machine learning).

Symbolists assert that intelligence can be reduced to symbols or rules that are based on high-level symbolic (i.e., human-readable) representations. A symbolic system takes physical patterns or symbols (e.g., chess pieces), combines them into structures using expressions (e.g., rules or logic of chess), and manipulates them using processes to produce new knowledge. A critical view of symbolic systems is that they seem appropriate for higher-level intelligence (e.g., reasoning) required to play chess but less appropriate for lower-level intelligence such as vision (an area where deep learning performs well). By themselves, symbolic systems prove to be brittle, but they were largely developed in an era with vastly less data and computational power than we have now.

Douglas Lenat, CEO of Cycorp, Inc., claimed "Intelligence is ten million rules." Lenat believed the time would come when a greatly expanded version of his software, Cyc, would underlie countless software applications. By 2017, he and his team had spent about two thousand person-years building approximately twenty-four million rules and assertions (not counting "facts"). Lenat has found purpose in the inelegance of his symbolic solution. It turns out that knowledge acquisition is time consuming and costly. In AI, the costly acquisition of knowledge has come to be known as the knowledge acquisition bottleneck. Lenat's story is told as a cautionary tale of what happens when you don't have a strong learning component like machine learning and have to create software by hand that is forever static.

Lenat's work does nonetheless highlight how much ambient knowledge there is in the world. His work suggests there is certainly problem-specific knowledge available for problem solving and that, maybe, we don't need to learn everything from scratch using machine learning. In other words, it is okay to know something about the problem you are solving. Heuristics, simple models, simple rules, and knowledge bases are ways of putting aspects of a problem back into a solution. I work with many clients who insist that their solution must be machine learning and that everything needs to be learned from scratch. The result is a data acquisition bottleneck or at least a playbook that produces the most expensive solution, not the cheapest, or even best, solution.

Ultimately, there is little love for symbolic representation by the artificial neural network tribe (i.e., connectionists). Connectionists are highly critical about the way symbolists work because they (connectionists) often consider describing something via a set of rules as just the tip of the iceberg. Arguments against symbol processing certainly show that human thinking does not consist solely of high-level symbol manipulation. However, this argument does not show that symbolic processing is useless, only that more than symbol processing is required for solving intelligence.

In his 1985 book, *Artificial Intelligence: The Very Idea*, John Haugeland coined the term "Good Old-Fashioned Artificial Intelligence" (GOFAI: pronounced goofy) to describe symbolic representation. While the popularity of sub-symbolic, or machine learning, has outpaced GOFAI it's not uncommon to improve performance by combining solutions together to create better solutions over machine learning alone. Combined learning includes symbolic reasoning (which further includes logic, expert knowledge, and knowledge bases containing context, facts, and rules), machine learning, and hybrid learning (the combination of various learning paradigms and learning architectures). As previously noted, real-world solutions are often sloppy and include a variety of partial solutions. Solutions will come from a variety of places, including data, rules, knowledge bases, various analytical families, and learning paradigms.

In fact, I frequently advocate for hybrid solutions to industrial partners for complex problems. Even for archetypical deep learning problem sets like computer vision there are many tools that can be layered together with less fashionable tools from traditional computer vision to improve performance. The reason to use hybrid approaches may be because you have small data where deep learning isn't by itself the best tool. Not to put too fine of a point on it, but an important part of problem solving is knowing when to use hybrid approaches and ultimately how to orchestrate partial solutions.

Specifically, GOFAI is often in the background—in one form or another—of many real-world solutions. So why is GOFAI still in so many real-world solutions? Well, GOFAI helps you get your solution out the door faster. Additionally, customers care about stable value propositions. Machine learning and statistical learning solutions are inherently probabilistic. Probabilistic solutions are limited by induction and have unstable value propositions. Therefore, having a more stable system operating in the background or foreground provides a more stable solution to customers. Regardless of any enthusiasm for combined learning and hybrid systems, it seems improbable that symbolic representation—even implemented flawlessly—would be the foundation of real AI or even the single missing piece.

The Streetlight Effect

Today, insiders acknowledge that there are missing pieces to achieving general intelligence, but no one can say with certainty what they are, how they fit together, or even if something like deep learning is the best starting point. Like Marcus, Pedro Domingos—a prominent insider and author of *The Master Algorithm*—advocates for an integrated, hybrid approach. He says that deep learning doesn't allow different pieces of knowledge to be composed in arbitrary ways like symbolist solutions do; nor does it evolve structure like evolutionary algorithms, properly handle uncertainty like Bayesian methods, or generalize to very different situations like analogical reasoning. Domingos suggests that combined learning is more comprehensive and necessary to make progress toward general intelligence.

Yann LeCun, chief AI scientist at Facebook and inventor of Convolutional Neural Networks, is advocating for integration as well. LeCun considers unsupervised learning to be a rather vexing issue on the path of AGI.[4] LeCun uses a cake metaphor for intelligence to describe supervised learning as the icing and reinforcement learning as the cherry, whereas unsupervised machine learning is the actual cake.[5] LeCun says we need to solve the unsupervised learning problem before we can even think of getting to true AI.

Reinforcement learning (RL), the cherry on the cake, is a learning paradigm that focuses on how so-called agents take action within an

---

4. Yann LeCun et al., "Gradient-based Learning Applied to Document Recognition," *Proceedings of the IEEE* 86, no. 11 (1998): 2278–324.

5. Yann LeCun, "Predictive Learning, NIPS 2016 | Yann LeCun, Facebook Research," August 23, 2017, www.youtube.com/watch?v=Ount2Y4qxQo&t=1072s.

environment to maximize a cumulative reward. For example, RL uses recursive interaction with the environment to fit a function that estimates the expected return on taking some random actions. By taking random actions, RL does not require any knowledge of the environment prior to learning. However, the agent is strictly associated with its environment. Imagine you want to teach an agent to play Mario Bros.[6] The agent can interact within the constraints of the game by choosing random actions such as run, jump, or duck, doing so thousands of times to achieve the goal of completing the stage. Unfortunately, agents are most often found in virtual environments rather than real-world ones, and some agent trained on Mario Bros would not be good for other problems that do not fit the problem characteristics of games. Therefore, an agent outside the original environment is useless, or not an agent at all.

The actual cake, the unsupervised machine learning, does not have the same demands on knowledge representation, such as logic and rules in symbolic solutions, or the specific demands on labeled data required for supervised machine learning. Unsupervised machine learning attempts to partition the data into largely homogenized groups or organize data into similar representations. More colloquially, think of birds of a feather flocking together. Structure is learned directly from the data and not defined in advance with classes and labels.[7]

An early, mainstream example of unsupervised machine learning is the famous Google cat.[8] In 2012, Google proved an artificial neural net-

6. Sebastian Heinz, "Using Reinforcement Learning to Play Super Mario Bros on NES Using TensorFlow," *Towards Data Science* in *Medium*, May 29, 2019, towardsdatascience .com/using-reinforcement-learning-to-play-super-mario-bros-on-nes-using-tensorflow -31281e35825.

7. There is no requirement for ground-truth because it would effectively be learned by the program. That said, these solutions are still very tricky to work with, and they often shift the work of defining the structure of the problem to after it gets solved but they do not remove the work.

8. A 2012 *New York Times* article ruefully poses the question, "How Many Computers to Identify a Cat? 16,000." John Markoff, "How many computers to identify a CAT? 16,000," *New York Times*, June, 25, 2012, https://www.nytimes.com/2012/06/26/technology/in-a -big-network-of-computers-evidence-of-machine-learning.html.

work with ten million randomly selected video thumbnails from YouTube. The algorithm was able to discover a variety of categories including human faces, which it predicted with an accuracy of 81.7 percent, and cats, which it predicted correctly 74.8 percent of the time.[9]

A more serious example of unsupervised learning is Google's word-2vec. Word2vec transforms words into numeric vectors to find word similarity and clustering. The basic premise behind word2vec is that words are collocated, and so, you can learn these dependencies directly from documents and represent words as numbers—meaning you can then do math on words.[10] A famous example that shows some of the incredible properties of word vectors is the concept of analogies. For example, the most famous example is the formula of "king" subtracting "man" and adding "woman" with the result of "queen."

Unsupervised machine learning may provide an important component for "real" AI—but the cake metaphor is just that, a metaphor. David Deutsch, a physicist at the University of Oxford and author of *The Beginning of Infinity*, says that using metaphors based on mistaken worldviews without first fully understanding AGI in detail "is like expecting skyscrapers to learn to fly if we build them tall enough."[11] In fact, artificial intelligence may not be an integration problem at all. After all, adding one, several, or hundreds of missing pieces to the puzzle is not important if the pieces are wrong or you're not building a puzzle at all.[12]

---

9. Quoc V. Le et al., "Building High-Level Features Using Large Scale Unsupervised Learning," *arXiv*, July 12, 2012, arXiv:1112.6209, https://arxiv.org/abs/1112.6209.

10. You can tell a lot by a name. For example, word2vec converts words to vectors, which is amazingly effective at communicating the scope of the effort. Tomas Mikolov et al., "Distributed Representations of Words and Phrases and their Compositionality," *arXiv*, October 16, 2013, arXiv:1310.4546, https://arxiv.org/abs/1310.4546. Moreover, names like artificial neural networks also communicate scope, which is neurobiological computing. Unfortunately, this name is less helpful for real-world problem solving.

11. D. Deutsch, "Creative Blocks: The Very Laws of Physics Imply That Artificial Intelligence Must Be Possible. What's Holding Us Up?" Aeon, October 3, 2012, https://aeon.co/essays/how-close-are-we-to-creating-artificial-intelligence.

12. P. Sweeney, "Sharpening the AI Problem," March 6, 2021, https://www.explainablestartup.com/2019/07/sharpening-the-ai-problem.html.

The journal *Nature Communications* recently took a strong position on the issue with a piece that argues that the entire enterprise of artificial intelligence research is based on a false assumption—namely that biological systems work because of some magical integration of unsupervised, supervised, and reinforcement learning. However, it's widely argued that the behavior of many, if not most, animals is almost completely hardwired at birth—as it must be if the animal is to have any hope of survival in a highly hostile environment.[13]

Essentially, the article argues that insiders often seek answers like the proverbial drunkard searching for their keys under a streetlight. Because of an observational bias known as the streetlight effect, they search where it's easiest to look, illustrated by a joke in which a police officer sees a drunk person searching for something under a streetlight. The officer asks the drunk what they've lost. The person says they lost their keys. So they both look under the streetlight together. After a few minutes, the police officer asks if they are sure they lost the keys here. The drunk replies, no, that they lost their keys in the park. The police officer asks why they're searching here, and the drunk replies, "This is where the light is."

This same point is illustrated in a 1988 revised edition of their seminal book titled *Perceptrons: An Introduction to Computational Geometry* where the authors, Marvin Minsky and Seymour Papert, wrote: "The movement of research interest between the poles of connectionist learning and symbolic reasoning may provide a fascinating subject for the sociology of science, but workers in those fields must understand that these poles are artificial simplifications. It can be most revealing to study (artificial) neural nets in their purest forms, or to do the same with elegant theories about formal reasoning. Such isolated studies often help in the disentangling of different types of mechanisms, insights and principles. But it never makes any sense to choose either of these two views as one's only model of the

13. A. M. Zador, "A Critique of Pure Learning and What Artificial Neural Networks Can Learn from Animal Brains," *Nature Communications* 10, no. 3770 (2019), https://doi.org /10.1038/s41467-019-11786-6.

mind. Both are partial and manifestly useful views of a reality of which science is still far from a comprehensive understanding."[14]

The *Nature Communications* article certainly highlights that insiders aren't going to solve intelligence with food metaphors or brain analogies alone. Perhaps symbolism could flourish again within new, unified paradigms or hybrid paradigms. The hybrid approach advocated by *Nature Communications* and Marcus may work somewhat like biological systems in the sense that there are numerous different systems at work together, including innate knowledge. How we can fuse the capabilities of machine learning techniques like deep learning with symbolic systems is still unclear, and the path to general intelligence is anyone's guess.

Paradigms in AI

Integration is certainly important in the real world and may be important for AI. As we have already discussed, real-world solutions are often an integrated mix of partial solutions. While it is unclear if intelligence is an integration problem, the tribal nature of scientific research makes integration difficult. Thomas Kuhn, author of the classic *Structure of Scientific Revolutions*, notes that science is like many other human professions, where the driver is not always shared ignorance used as progress toward some absolute measure of truth, but shaped by shared knowledge that researchers in a field tacitly and unquestionably believe in. Kuhn's somewhat cynical view on group preferences is what he calls "paradigms."

Paradigms explain why scientists tend to be good for one revolution and not many revolutions. The reason is that an insider may not gain from another revolution where any attempt may subvert the trust that they have with their paradigm. In his book *Scientific Autobiography*, Nobel Laureate Max Planck coined the so-called Planck's principle, which states that scientific change does not occur because individual scientists change

---

14. M. Minsky, S. Papert, and L. Bottou, *Perceptrons: An Introduction to Computational Geometry* (Cambridge, MA: MIT Press, 2017).

their mind, but rather successive generations of scientists have different views.[15, 16] Informally, this is paraphrased as "Science progresses one
funeral at a time."

We ought to accept that solving intelligence is an intractably hard problem. Lacking explicit guidance on solving something like AGI is understandable. At the time of this writing, Marcus does not clearly outline how
some integration would be designed, but he is willing to admit it. Author
Joanne Gaudet notes that knowledge mobilization is mobilizing what is
known. Ignorance mobilization is mobilizing what is unknown.[17] If we are
mobilizing ignorance, we are working to develop knowledge about what is
unknown. Thus, discovery of new knowledge starts with ignorance and a
shared interest in a question about something that is unknown.

Of course, while Marcus is sincere in his critique, he may be missing
the point that paradigms don't change by arguing. Kuhn and Planck highlight that a paradigm changes only when it is replaced. As Yale University
professor Drew McDermott noted, a common folly in research is to suppose that having identified shortcomings is equivalent to having actually
solved shortcomings.[18] Despite his assiduous critique, Marcus may have
overlooked that naming a demon isn't the same thing as destroying it. In
sum, it is odd to criticize Marcus for dodging how to practically implement

15. Max K. Planck, *Scientific Autobiography and Other Papers* (New York: Philosophical
Library, 1950), 23–24.

16. Geoffrey Hinton seemed sensitive to Planck's principle when he encouraged researchers to be deeply suspicious about back-propagation being the future of AI. Hinton added
that back-propagation should be thrown away and we should start over. Hinton under-
scored that insiders are not driven by partial success and that their goals are academic
in nature, not economic. *See* Steve LeVine, "Artificial Intelligence Pioneer Says We Need
to Start Over," *Axios*, December 15, 2017, www.axios.com/artificial-intelligence-pioneer
-says-we-need-to-start-over-1513305524-f619efbd-9db0-4947-a9b2-7a4c310a28fe.html.

17. J. Gaudet, "It Takes Two to Tango: Knowledge Mobilization and Ignorance Mobilization in Science Research and Innovation," *Prometheus* 31, no. 3 (2013): 169–87, http://dx
.doi.org/10.1080/08109028.2013.847604.

18. McDermott precisely wrote, "A common idiocy in AI research is to suppose that having identified the shortcomings of Version I of a program is equivalent to having written Version II." Drew McDermott, "Artificial Intelligence Meets Natural Stupidity," *ACM
SIGART Bulletin* 57 (April 1976): 4–9, https://doi.org/10.1145/1045339.1045340.

his hybrid ideas.[19] It is also odd, however, that Marcus seems to want others to agree and perhaps even fix anomalies in order to create a new paradigm that would benefit him.

Of course, by marginalizing and suppressing commentary, the connectionists implicitly claim that if others like Marcus can see a problem, then the problems can also be seen by their solutions. This is wishful thinking. Deep learning and machine learning do not possess any awareness of their ignorance, share in any of the responsibility from their mistakes, or recognize any of their own shortcomings.

What is particularly odd about these debates—and perhaps unique to AI—is that insiders are not only disagreeing on the best solution (which is rather common in branches of science and engineering), but also on the problem.[20] One prominent figure in the AI Twitter debate of 2018 was Turing Award winner Yann LeCun. Marcus and LeCun are insiders in different paradigms nested within the same larger paradigm. In other words, both Marcus and LeCun are working in different parts of the same whole. LeCun understands the problem to be that humans are not hardwired thus requiring better learning. He and the connectionists believe that AI will require deep learning because deep learning kind of looks like the brain. Marcus argues for more innate machinery producing better reasoning. Marcus and the symbolists believe that AI will require something more than deep learning because deep learning is not the brain.

The point is that they are solution arguing, where even those arguments are aligned with paradigms, not necessarily with truth, and certainly not with any business goal in mind. Ultimately, a business needs to maintain a balance while spanning boundaries and paradigms. Balance means there is no score keeping. There is no good reason to show support for any paradigm that isn't your own. The reason is important, as we must remember that paradigms have their own movement, and their own

---

19. Theoreticians pontificate, but to assume responsibility. Businesses operate in the opposite mode where action is required and well said is not the same thing as well done.

20. Insiders struggle with problem definition as well.

understanding of the problem, even though their overriding goal, which is to solve intelligence, is the same. Ultimately, we must create our own goals and solve our own problems.

The Forest for the Trees

Solution arguing is part of the orthodoxy of AI. It may have started in 1969 when Minsky and Papert wrote their damning critique on perceptrons. More recently (December 2019), Marcus debated Turing Award winner Yoshua Bengio on the topic of symbolic representation in deep learning.[21] These debates are common and not new to the field of AI. Insiders are passionate, and they love to argue about solutions.

Marcus's arguments, while important, are not exactly novel. In fact, the LeCun-Marcus divide may just be old wine in a new bottle as debates about what AI "ought" to be are as old as AI itself. Examples include Minsky versus Rosenblatt, Chomsky versus Skinner, Carey versus McClelland, Smolensky versus Fodor and Pylyshyn, Lakoff versus Pinker, and Bates versus Pinker. The list goes on and on.

During the late 1950s and early 1960s, Rosenblatt and Minsky debated on the floors of scientific conferences about the value of biologically inspired computation. Sound familiar? Rosenblatt argued that perceptrons could do almost anything, while Minsky countered that they could do little. Again, sound familiar?

While these generally stimulating debates may be great intellectual fodder and water-cooler conversation, they are often bad for business. The issue is when these debates spill over to adjacent regimes and business-minded people confuse them as important. They are not. Although Marcus's critique is good for the discipline of artificial intelligence, it also reinforces a culture that is a bad fit for businesses.

---

21. Matt Taylor et al., directors, "Gary Marcus vs Yoshua Bengio 2019 University of Montreal," YouTube, Numenta, December 24, 2019, www.youtube.com/watch?v=aQQrpG3VJOk.

In other words, the lack of answers is what makes the scientific discipline of artificial intelligence interesting, and often, robust. However, it is not difficult to imagine that arguing about solutions and whether these solutions have satisfied the external goals created for them creates misalignments. We must keep in mind that throughout his critique of deep learning Marcus maintains that the *most important* question is, "Can it [deep learning] solve general intelligence?" More broadly, this can be interpreted as, "Can machine learning solve general intelligence?" It can't and certainly hasn't.

Whether deep learning can solve intelligence or solve all problems is not very interesting to business. Ultimately, what a solution *is* and what it *ought* to be should not be determined by someone who's not part of your business. At this point, we should recognize that the discussion around true, real, strong, general, or complete AI is a normative argument that ignores the way things are. Normative arguments are constructed through the lens of value judgments. A value judgment is an assessment of something as good or bad in terms of someone else's goals, values, standards, and priorities. Value-based judgments and normative arguments provide the undercurrent to these debates. Ultimately, value judgments allow people to marginalize solutions such as deep learning, machine learning, or symbolic representation because these solutions do not meet their goals, values, and standards.

The distinction between normative and descriptive claims is closely drawn from philosopher David Hume's "is-ought problem." The is-ought problem states that many people will make claims about what *ought* to be, based on statements about what *is*. However, the gap between "is" and "ought" statements[22] renders "ought" statements dubious because what something is cannot be what it ought to be. Or, stated somewhat tautologically, it is what it is.

---

22. When combined with Hume's fork, which is the idea that all items of knowledge are based either on logic and definitions or observation.

The is-ought problem holds that what AI ought to be cannot be known to AI. Ignoring that this is an is-ought argument, you realize that today's solutions are the only solutions we have. They are useful (i.e., economically viable), yet fallible, and the only way to know if fallible is useful is to first understand the problem you need to solve. That problem is never anything as abstract as intelligence.

Remember that insiders seek epistemological discoveries, not economic ones. The more epistemological a pursuit is, the less likely it is to become something that could be turned into a business. Entrepreneur, venture capitalist, and author Paul Graham discusses the value of problems at length and explains that good business ideas are unlikely to come from scholars who write and defend dissertations.[23]

The reason is that the subset of ideas that count as "research" is so narrow that it's unlikely to satisfy academic constraints and also satisfy the orthogonal constraints of business. The incentives for success in the academic world are not consistent with what it takes to start and grow a business. Ultimately, business pursuits are much more complicated than academic ones. Managers ought to acknowledge that solving intelligence is not likely your goal, and in many ways it's oppressive to problem solving. AGI may be possible, but it is not desirable as a business goal.[24]

Fundamentally, these debates establish new specifications of intelligence and new arguments on how intelligence must be solved. Although Marcus's critique eases the heavy-handed treatment of deep learning by many in the media as a solution to everything, he adds another heavier hand. Although it's good to have knowledge of the limitations of deep

---

23. Paul Graham, "How to Get Startup Ideas," paulgraham.com, November 2012, www .paulgraham.com/startupideas.html.

24. Moreover, there is no existential risk of betting against general intelligence. Roko's Basilisk is a thought experiment about the potential risks involved in developing so-called true artificial intelligence. The premise is that an all-powerful artificial super-intelligence could retroactively punish those who did not help bring about its existence. Roko's Basilisk resembles a futurist version of Pascal's Wager in that it suggests people should weigh the possible infinite gain and infinite punishment for accepting or rejecting the creation of artificial super-intelligence. You don't.

learning and machine learning, knowing how intelligence ought to be solved is not. Business doesn't need to secure or adopt a metaphysical vision of AI's supposed destiny. In most cases, businesses need only marginal improvements to gain competitive advantages. Imagine what a 10 to 20 percent improvement would mean for your customers, especially if those gains were unrealized by your competitors, who were stuck arguing about how good a solution ought to be using someone else's goal.

Calling a solution "AI" may be in vogue, but there is always the risk that we'll spend more time arguing about the name than arguing about the impact some imperfect solution might have on a problem. Machine learning, deep learning, symbolic reasoning, combined learning, and hybrid systems are all just imperfect tools. When solving problems, they mean much less than the problem itself. The forest for the trees.

HUMAN MEASURING
STICKS

When a measure becomes a target, it
ceases to be a good measure.
**—Marilyn Strathern, paraphrasing
Goodhart's Law**

anagement philosopher Peter Drucker famously said that "you can't manage what you can't measure" and "if you can't measure it, you can't improve it." What Drucker means is that you can't generally know whether you're successful unless your goals are known and progress is tracked against those goals. It's much easier to track and produce more desirable outcomes with clearly established goals and metrics.

To be sure, measuring progress toward success demands some sort of measuring stick. Businesses are filled with measuring sticks such as gross margin, sales, churn, customer acquisition costs, revenue, customer lifetime value, net promoter score, and many, many more. Measuring sticks support various aspects of an organization, including finance, marketing, sector performance, customer requirements, customer satisfaction, and forecasting. However, there are two issues with Drucker's famous words. First, he never actually said it.[1] Second, while measurement is important, it is not a substitute for good management or problem solving.

Without clear business objectives, metrics, and goals may default to the factory setting of artificial intelligence in which the standard of success is the human standard. For business, however, this is the wrong standard and represents the wrong metric. Whereas insiders can judge their own goals, many real-world projects run the risk of falling into the *intelligence trap*. The intelligence trap occurs when solutions are viewed centrally, from the point of view of an abstract human or an abstract idea of intelligence, using an abstract human measuring stick. You might be falling into the intelligence trap if you are creating a solution that simply measures itself against yourself. In such a scenario, you're not looking hard enough for a customer. Instead of managing an actual problem, you may be simply measuring someone else's goals.

A human measuring stick is a poor metric for business because it doesn't align with a specific problem, customer, or a tangible value proposition that includes problem-specific information. We must understand that for the insider everyone is a customer because everyone is human, and all humans have intelligence. Consequently, abstract human measuring sticks make sense. However, your customers are never just human. When they are defined in such a general way there's no way to discern one customer from another or ourselves from our customers. Businesses have to know

---

1. https://www.drucker.institute/thedx/measurement-myopia/.

more about a customer than the simple fact that they are human. If you don't, you'll have no idea how to impact their lives.[2]

Consider how artificial intelligence research generally seeks to "beat" a physician's ability to predict a disease.[3] In fact, perform an internet search for "AI beats human," and you will find a trove of examples of so-called AI beating humans at games, art, law, and everything in between.[4] Beating a physician, however, is not a useful metric for the physician. More specifically, human measuring sticks do not translate to any real-world impact for the physician. This fundamental mismatch between the way success is measured and the way doctors work was a central critique on IBM's Watson Health.[5]

However, what works for *Jeopardy!* may not work for real problems.[6]

This type of metric is designed for vanity, not real-world value. Instead, consider the target of predicting normal heart function.[7] This is a metric

---

2. The "jobs to be done" theory, by Clayton Christensen, asserts that people buy products and services because they are trying to make progress in their jobs. So, it's your job to know enough about your customer's job to know how they might define progress. Knowing if someone is human or even having demographic information will not drive success. Clayton M. Christensen et al., *Competing against Luck: The Story of Innovation and Customer Choice* (New York: Harper Business, 2016).

3. Search Google for "AI beats doctors," and you can see for yourself all the clickbait.

4. J. Miley, "11 Times AI Beat Humans at Games, Art, Law and Everything in Between," *Interesting Engineering*, May 2, 2018, https://interestingengineering.com/11-times-ai-beat -humans-at-games-art-law-and-everything-in-between.

5. Consider the project Watson for Oncology and how its lack of problem comprehension (and how real doctors work) doomed it to failure and sent $62 million down the drain. IBM commercials at this time were giving strong signals that IBM already had intelligence that was indistinguishable from human intelligence, https://www.youtube.com /watch?v=8xYvwcnHn9k. Paired with some FOMO, misleading demos, and the "nobody gets fired for buying IBM effect," Watson garnered a lot of buzz at the time. However, of the nearly fifty partnership announcements made by IBM, very few have led to commercial products. E. Strickland, "How IBM Watson Overpromised and Underdelivered on AI Health Care," *IEEE Spectrum*, April 2, 2019, https://spectrum.ieee.org/biomedical/diagnostics /how-ibm-watson-overpromised-and-underdelivered-on-ai-health-care.

6. G. Marcus, "DeepMind's Losses and the Future of Artificial Intelligence," *Wired*, August 8, 2019, https://www.wired.com/story/deepminds-losses-future-artificial-intelligence/.

7. Wiebke Toussaint et al., "Design Considerations for High Impact, Automated Echocardiogram Analysis," *arXiv*, June 2020, arXiv:2006.06292.

that would actually save cardiologists time by identifying patients who do not need their expertise. In other words, we can help customers by using problem-specific information rather than just trying to beat medical professionals. This is the difference between solving intelligence—and marginalizing flesh-and-blood people—and solving problems the world presents to intelligence, thus *helping* flesh-and-blood people.

Threshold Guessing

Because the central goal of artificial intelligence is solving intelligence, insiders use human measuring sticks and human thresholds to define and track progress.[8] For example, contests determine winners by comparing solutions that exceed "human-level" thresholds. However, these thresholds aren't what you might expect. Thresholds are evaluated using the so-called top-five error.[9] This means the correct classification for an object can be any one of a solution's top five predictions.

Professor at the Santa Fe Institute, Melanie Mitchell, writes in her book, *Artificial Intelligence: A Guide for Thinking Humans*, "if, given an image of a basketball, the machine outputs 'croquet ball,' 'bikini,' 'warthog,' 'basketball,' and 'moving van' in that order, it is considered correct." If a picture shows more than one object, it's useful to know whether the machine is detecting any of them. However, it also weakens the claim of extreme performance. The top-one accuracy (still part of the less than realistic "closed" problem) was 82 percent in 2017.[10] This figure is not often

---

8. When we're dissatisfied with human measuring sticks or when we realize the loftiness they create, we may create equally flawed intelligence measuring sticks, like those programs that can pass a Turing test or Super-Turing test; reflect the intelligence of a third grader, puppy, chimp, cockroach, rat, or cat; or exhibit Nobel-*like* intelligence and, of course, superhuman-intelligence.

9. The top-1 accuracy is reported more frequently in recent years.

10. Hang Zhang et al., "ResNeSt: Split-Attention Networks," *arXiv*, December 30, 2020, arXiv:2004.08955. *See also* https://gluon-cv.mxnet.io/model_zoo/classification.html#id250 for code and visualization.

published and is well below the 95-percent threshold metric for humans, which itself is a statement about human performance that is quite flimsy.

These thresholds are more of a guess, which is anthropomorphized post hoc. Instead of developing a theory of intelligence or establishing actual threshold of human performance, insiders guess at thresholds. For example, ImageNet compares solution performance with the 95-percent accuracy threshold, which is meant to represent the human standard of performance for the task of object recognition. This threshold, however, was created by one person who studied a set of five hundred labeled images and then sorted unlabeled images into categories. The result was a 5-percent error rate.[11] That is, using the top-5 accuracy as the metric, 5 percent of the single-human curation was wrong.[12] That figure of 5 percent now gets used as the mark of human performance despite the fact that many of the errors did not arise from a large sample of people who were clueless about the objects they were seeing but from one person who didn't recall some of the exact esoteric categories used in ImageNet.[13]

Practically speaking, thresholds are meant to benchmark solutions by stressing the differences and similarities between solutions. Consequently, human thresholds are meant to stress the difference and similarities between human and computer performance. However, what is actually used as thresholds do not truly represent the bounds of human ability.

---

11. Imagine if this was true for driving where you, and everyone around you, were only able to recognize objects 95 percent of the time. While 95 percent performance is passable for image hosting sites like Google Photos you could also upset your customers and have them resent you.

12. Andrej Karpathy, AI director at Tesla, studied a labeled set of 500 images as his "training data" and then sorted 1,500 unlabeled images into categories. With top-5 accuracy as the metric, he got 5 percent wrong. And now, that figure gets used (*see* https://twitter.com/karpathy/status/786356525010780165?lang=en) as the mark of human performance. Karpathy has written (*see* karpathy.github.io/2014/09/02/what-i-learned-from-competing-against-a-convnet-on-imagenet/) that about one-fourth of the errors made by him and the other test subject arose not because they were clueless about the pictures they were seeing, but because they didn't know or didn't recall some of the exact labels used in ImageNet.

13. Andrej Karpathy, "What I Learned from Competing against a ConvNet on ImageNet," *Andrej Karpathy* (blog), September 2, 2014, karpathy.github.io/2014/09/02/what-i-learned-from-competing-against-a-convnet-on-imagenet/.

More often, these thresholds reflect the idiosyncratic nature of benchmark data sets and the poor consensus of human labelers.[14] As mentioned in chapter 3 (see "Benchmarks and Contests"), contests tend to characterize benchmark data sets. So, too, do thresholds, which tend to tell us more about the categories of benchmark data sets or the arbitrariness of thresholds than about machine performance or actual human performance.[15]

Social Comparison Theory (SCT) may explain why insiders evaluate their solutions by comparing them to themselves. SCT, initially proposed by social psychologist Leon Festinger, centers on the belief that there's a drive within individuals to gain accurate self-evaluations. SCT explains why targets often share distinctive characteristics with evaluators as it helps ensure accurate self-evaluations. Needless to say, self-evaluations do not result in a self-referencing solution that mimics natural intelligence, self-awareness, self-reflection, or any sort of mental imagery. In other words, human measuring sticks do not make a solution human; they reflect only a goal and not always accurately.

Take, for example, the Stanford Question Answering Dataset (SQuAD). SQuAD is a benchmark data set. It is to language what ImageNet is to pictures.[16] Specifically, SQuAD is a question-answering data set compiled by workers hired through Mechanical Turk. Amazon launched Mechanical Turk in 2005. Mechanical Turk recruited human labor to perform simple tasks for nominal fees. But do low-wage workers with few incentives to do

14. "Excavating AI" by Kate Crawford and Trevor Paglen documents how strange some of the consensus labeling actually is. For example, a woman smiling in a bikini is labeled a "slattern, slut, slovenly woman, trollop." A young man drinking beer is categorized as an "alcoholic, alky, dipsomaniac, boozer, lush, soaker, souse." A child wearing sunglasses is classified as a "failure, loser, non-starter, unsuccessful person." You're looking at the "person" category in a data set called ImageNet, one of the most widely used training sets for machine learning. Kate Crawford and Trevor Paglen, "Excavating AI: The Politics of Training Sets for Machine Learning," September 19, 2019, https://excavating.ai.

15. As the maxim goes: what gets measured, gets manipulated. Nassim Taleb recently suggested in a paper on p-hacking that metrics are stochastic and what is stochastic will be hacked. Nassim Nicholas Taleb, "A Short Note on P-Value Hacking," 2008, https://arxiv.org/abs/1603.07532.

16. Fei-Fei Li used Amazon Mechanical Turk for ImageNet.

well performing such a task without constraints truly represent human ability?[17] Do benchmark data sets or human thresholds truly represent the real-world context for real-world problem solving? Real-world constraints are vital for real-world problem solving. For example, object recognition on a benchmark data set is vastly different than object recognition for self-driving cars, where the context of objects is literally changing, and the context of the problem also includes flesh-and-blood people at risk.

The abstraction of some average human ignores what often makes technology solutions so interesting. Consider x.ai, which connects email and calendars to coordinate the best time to meet. Founded in 2014, x.ai is a productivity tool "powered by artificial intelligence." This solution isn't very interesting for average email users, but it may be very interesting for power email users managing several calendars. If we don't understand our customer, who in early stages of customer acquisition are often extreme users, then what can we understand from abstract human measuring sticks?

Besides, for most real-world problems the human measuring stick is a standard too high. In truth, it's a great way to needlessly demand a complex solution to a problem. For example, a computer vision problem may need only five or ten thousand observations per category to get acceptable performance on a task measured against a specific problem, whereas human-level performance may need one million or even ten million observations per category. Most companies—small and large—will have a desire for capital efficiencies that would preclude some desire for maximum performance thresholds.

The true point is more elusive. That is, for some problems you will not know what threshold you need to satisfy or exceed. Consequently, the optimal solution will not be known in advance. In these circumstances, maximum performance thresholds should be avoided. Just as software development products use minimum viable products to collect feedback

---

17. Gary Marcus, "An Epidemic of AI Misinformation," *The Gradient*, November 30, 2019, thegradient.pub/an-epidemic-of-ai-misinformation/.

from initial customers, so too should AI solutions target minimum algorithmic performance required by early adopters.

The ELIZA Effect

The best application of human measuring sticks may be an expectation for
the type of performance we can anticipate from computers given a similar
data set. However, this expectation has at least two serious pitfalls. One is
that we cannot prioritize a single metric or threshold when problem solving. Insiders try to do this, but we should be careful when following them
down this path.

Another issue with human measuring sticks is that they lead to over-
attribution and general confusion. When we over-attribute solution performance to natural intelligence, either ex ante or ex post, we lose agency.
The loss of agency is especially true when we project human characteristics
to our solutions. It's a mistake to presume that this transfer of authority
involves a simultaneous absolution of responsibility. Solution performance
may at times overlap with human function, but it's also important to know
that performance is not actual human function. Said differently, AI, in its
current form, is a servant, or supplement, for human intelligence, not a
master, or substitute, for human intelligence.

The danger of conflating human measuring sticks with human intelligence is that this type of comparison is never justified and ultimately
results in a type of measurement error, or the difference between a measured quantity and its true value. We run the risk of mistaking an overlap in function with an overlap in intelligence, which is spurious. In other
words, solutions do not understand the problem, the output, or the context
of the problem, nor do they share in any responsibility, like humans do.

Machine learning solutions will make strange mistakes that are difficult to debug. These mistakes will not be psychologically reasonable. They
will fail in ways that are counterintuitive and nonobvious. This can be due
to anything from skewed training data to unexpected interpretations of
data during training. Furthermore, when machine learning is incorporated

into our products, the interactions and feedback loops will be complicated, making it difficult to predict and test all possible situations. These challenges require product teams to spend a lot of time figuring out what these solutions are doing and how to improve them.

Joseph Weizenbaum, professor of computer science at the MIT Artificial Intelligence Laboratory, created ELIZA in the mid-1960s. ELIZA was an early chatbot fashioned after a psychotherapist. Weizenbaum reasoned psychotherapists are an easy human form to imitate because psychotherapy is inherently opaque.[18] What surprised Weizenbaum was that early users of ELIZA were convinced of ELIZA's intelligence and understanding, despite Weizenbaum's claims to the contrary.[19] The ELIZA effect has since come to be understood as the eagerness to project human intelligence onto inanimate objects. That is, we see AI where there is no AI.[20]

The ELIZA effect explains our tendency to unconsciously assume computer behaviors are analogous to human behaviors. It explains the very human tendency to attach meaning to things that our solutions never put there.[21] The removal of the human from measuring sticks will not magically solve all the unique ways solutions fail, but it will help us anticipate failures by removing the rose-colored, anthropomorphic glasses.

---

18. Specifically Rogerian psychotherapy, developed by Carl Rogers, which is still widely used today.

19. Artificial intelligence may be possible, but Weizenbaum argued that we should never allow computers to make important decisions. Weizenbaum makes a crucial distinction between deciding and choosing. He concludes that comprehensive judgment involves both mathematical and non-mathematical factors including emotions. J. Weizenbaum, *Computer Power and Human Reason: Judgment to Calculation* (San Francisco: W.H. Freeman, 1976).

20. Almost twenty years after ELIZA, Douglas Hofstadter made this mistake. Hofstadter was fooled by "Nicolai." Nicolai was said to be developed by the US Army, but it was actually a prank by some of his students. Hofstadter notes in his 1985 book how amazed he was at how much intelligence he was willing to accept. Douglas R. Hofstadter, *Metamagical Themas* (New York: Basic Books, 1985), 520, Kindle edition.

21. For example, *see* Amanda Kooser, "NASA Mars Rover Opportunity Earns Heartbreaking Eulogies: 'I'm Crying,'" *CNET*, February 13, 2019, www.cnet.com/news/nasa -mars-rover-opportunity-earns-heartbreaking-eulogies-im-crying/.

When a study reveals that changing an image of a lion in a way that is imperceptible to humans can cause a machine learning algorithm to predict class assignment of the image as something else entirely—mislabeling a lion as a library, for example—we ought to be aware of the brittleness. But we shouldn't be surprised. Machines will not fail gracefully; they will fail silently. Kelly's law states that old technology fails frequently but in a reliable way. New technology fails less often, but when it does, it fails in an unexpectedly new way we are not prepared for. In short, the newer the technology, the more novel the modes of failure.

Researchers Anh Nguyen, Jason Yosinski, and Jeff Clune outline how easy it is to produce images that are completely unrecognizable to humans but that state-of-the-art machine learning believes to be recognizable objects with 99.99 percent confidence (e.g., labeling with certainty that white-noise static is a lion).[22] The issue is that when a program is wrong, it's often wrong in ways that no human would ever be. No human would mistake a picture of a baby holding a toothbrush for a baseball bat.[23] It's important to consider where and how we would use these solutions.

Another serious challenge for computer vision is an oversensitivity to context (meaning an artificial neural network is focusing on the background of the picture rather than the foreground). Consider the effect of Photoshopping a guitar into a picture of a monkey in the jungle.[24] This causes machine learning to misidentify a monkey as a human and also to misinterpret a guitar as a bird, presumably because monkeys are less likely than humans to carry a guitar and birds are more likely than

22. Anh Nguyen, Jason Yosinski, and Jeff Clune, "Deep Neural Networks Are Easily Fooled: High Confidence Predictions for Unrecognizable Images," *arXiv*, December 2014, arXiv: 1412.1897, http://arxiv.org/abs/1412.1897.

23. Neil Savage, "Seeing More Clearly," *Communications of the ACM* 59, no. 1 (January 2016): 20–22, cacm.acm.org/magazines/2016/1/195740-seeing-more-clearly/abstract.

24. Alan L. Yuille and Chenxi Liu, "Deep Nets: What Have They Ever Done for Vision?" *arXiv*, May 2018, arXiv:1805.04025, http://arxiv.org/abs/1805.04025.

guitars to be in a jungle.[25] Other examples of oversensitivity to context include slight changes in object position. This can affect an object's perceived identity according to an object detector, as well as that of other objects in the image. For example, Amir Rosenfeld, researcher at York University in Toronto, as well as Richard Zemel and John Tsotsos of the University of Toronto, put an elephant in a room—a process called object transplanting—to trick an algorithm.[26]

Testing oversensitivity is accomplished by introducing something incongruous into the scene. It's a way to probe algorithms and their vulnerabilities. In recent years there have been a slew of attempts, known as "adversarial attacks," in which researchers contrive scenes to make machine learning fail. Adversarial data sets are data sets designed to make models predict erroneously. Think of it as a way to stress test your solutions.

In one experiment, LabSix tricked an artificial neural network into mistaking a turtle for a rifle.[27] In another, researchers "attacked" (i.e., tested) an artificial neural network by placing an image of a psychedelically-colored toaster alongside ordinary objects like a banana.[28] As for the elephant itself, an artificial neural network sometimes called an elephant a sheep, and sometimes it overlooked the elephant completely. Conversely, humans notice elephants in the room, both real elephants as well as the metaphorical idiom.

Rosenfeld explains that today's best machine learning algorithm for vision problems works in a *feed-forward* manner. This means that information flows through them in one direction. They start with an input of fine-grained pixels from an image, then detect the curves, shapes, and scenes,

25. Jianyu Wang et al., "Visual Concepts and Compositional Voting," *Annals of Mathematical Sciences and Applications* 2, no. 3 (2018): 4.

26. Amir Rosenfeld, Richard Zemel, and John K. Tsotsos, "The Elephant in the Room," *arXiv*, August 2018, arXiv:1808.03305,http://arxiv.org/abs/1808.03305.

27. Anish Athalye et al., "Fooling Neural Networks in the Physical World," *Labsix*, October 31, 2017, www.labsix.org/physical-objects-that-fool-neural-nets/.

28. Tom B. Brown et al., "Adversarial Patch," *arXiv*, December 2017, http://arxiv.org/abs/1712.09665.

with the network making its best guess about what it's seeing at each step along the way.[29] As a consequence, errant observations early in the process often end up contaminating the end of the process, when the solution pools together everything it "knows" to make a prediction about what it's "looking" at. These kludgy statistical approximations have no comprehensible, or defensible, understanding of the actual objects in the photos. Rosenfeld, Tsotsos, and Zemel are developing a theory called selective tuning, which explains these features of visual cognition.[30]

Humans approach the problem differently. Kevin Hartnett, senior writer at *Quanta Magazine*, asks us to imagine we're given a very brief glimpse of an image containing a circle and a square. One of them is colored blue, the other red. Afterward you're asked to name the color of the square. With only a single glance to go on, we're likely to confuse the colors of the two shapes. We're also likely to recognize that we're confused and take another look. Critically, when we take that second look, we know to focus our attention on just the color of the square. Having the capacity to reflect and check where we might have erred, we understand that our visual system lacks the right answer.[31]

Designing artificial perception programs is very hard, partly because we understand some of the ways that human vision works—like seeing edges and hierarchies—but not all.[32] And then there's the complexity inherent in

---

29. This is pretty accurate, though the hidden, or intermittent layers, are difficult to interpret.

30. J. K. Tsotsos, "Analyzing Vision at the Complexity Level," *Behavioral and Brain Sciences* 13, no. 3 (1990): 423–45.

31. Eric Nyquist, "Machine Learning Confronts the Elephant in the Room," *Quanta Magazine*, www.quantamagazine.org/machine-learning-confronts-the-elephant-in-the-room-20180920/.

32. Patricia Kitcher, "Marr's Computational Theory of Vision," *Philosophy of Science* 55, no. 1 (1988): 1–24, www.jstor.org/stable/187817; K. Fukushima, "Neocognitron: A Self-Organizing Neural Network Model for a Mechanism of Pattern Recognition Unaffected by Shift in Position," *Biological Cybernetics* 36 (1980): 193–202, https://doi.org/10.1007/BF00344251; D. H. Hubel and T. N. Wiesel, "Receptive Fields of Single Neurons in the Cat's Striate Cortex," *Journal of Physiology* 148 (1959): 574–91, http://jp.physoc.org/content/148/3/574.full.pdf+html; Cade Metz, "Jeff Hawkins Is Finally Ready to Explain

the visual world itself. The fact that speech recognition and object recognition have not been solved in any comprehensive manner should not shock us. Consider that an artificial vision system must be able to see in any of an infinite number of scenes and still extract something meaningful.

Psychologists Kurt Gray at the University of North Carolina and Daniel Wegner of Harvard University write that the high cost of failing to detect agents and the low cost of wrongly detecting them suggest that humans possess a hyperactive agent detection device or a cognitive module that readily ascribes events in the environment to the behavior of agents.[33] The hyperactive agent detection device suggests why humans see faces in the clouds but never clouds in faces. It's because we have special cognitive modules for face detection that have evolved for millions of years.[34] In other words, the human object detector is on a hair trigger, and it makes almost all of its mistakes in one direction—false positives (i.e., seeing a face when no real face is present), rather than false negatives (i.e., failing to see a face that is really present).[35] Said differently, a human may see a cloud that looks like a face, whereas AI may *see* a face that looks like a cloud.

Human cognition has developed and evolved over six million years to save our lives. The mind is the ultimate simulation algorithm, shuffling data to realize new outcomes in order to acquire new knowledge not for accuracy but for survival.[36] Evolution and human cognition are meant to improve our lives and our chance of survival. Human thinking is practical,

---

His Brain Research," *The New York Times*, October 14, 2018, www.nytimes.com/2018/10/14/technology/jeff-hawkins-brain-research.html.

33. Kurt Gray and Daniel Wegner, "Blaming God for Our Pain: Human Suffering and the Divine Mind," *Personality and Social Psychology Review* 14, no. 1 (November 2009): 7–16, doi:10.1177/1088868309350299.

34. Stewart Guthrie, *Faces in the Clouds: A New Theory of Religion* (New York: Oxford University Press, 1993), 110.

35. Jonathan Haidt, *The Righteous Mind: Why Good People Are Divided by Politics and Religion* (New York: Vintage Books, 2012), 292.

36. Another reason not to use the human standard. The reason is that human intelligence has evolved for survival and is not purely functional.

not ontological or epistemological. Being right is valuable, but it isn't the ultimate goal.

Even when we're relaxing, our brain is continuously combining and recombining information—some of which does not exist—to imagine the future. Researchers were surprised to find this process even after scanning the brains of people performing specific tasks such as mental arithmetic. "Whenever there was a break in the task there were sudden shifts to activity in the brain's so-called default circuit, which is used to imagine the future or retouch the past."[37] This discovery explains what happens when our mind wanders during a task; its simulating future possibilities, many of which are unsupported by any real data.

These simulations are what enable humans to respond so quickly to complex and difficult problems. They also explain why machines, which lack this understanding, are so inadequate at responding correctly to unexpected situations unsupported by data. What may feel to us like a primitive intuition—a gut feeling—is made possible by previous simulations. To minimize error machine learning merely fits a function to data to find patterns that it doesn't understand, based only on data it has processed. Some solutions can fit increasingly complex functions to complex data sets but are still limited by the data they have seen. So their algorithms lack any comprehension and have trouble extrapolating to unseen data that falls outside of the distribution on which they were trained.

All of this isn't to suggest that evolution doesn't play tricks on natural intelligence. Consider the following examples of ambiguous images that can form two separate pictures: a rabbit and a duck; the famous Pac-Man configuration popularized by Gaetano Kanizsa;[38] the Müller-Lyer illusion,

---

37. Martin Seligman and John Tierney writing in a 2017 *New York Times* article ("We Aren't Built to Live in the Moment," https://www.nytimes.com/2017/05/19/opinion /sunday/why-the-future-is-always-on-your-mind.html) about the research of Randy L. Buckner, Jessica R. Andrews-Hanna, and Daniel Schacter. R. L. Buckner et al., "The Brain's Default Network" in *The Year in Cognitive Neuroscience 2008* 1124, no. 1 (2008): 1–38, https://doi.org/10.1196/annals.1440.011.

38. Gaetano Kanizsa, "Subjective Countours," *Scientific American* 234 (1976): 48–52.

consisting of a stylized arrow; or the moon illusion. The moon illusion might be the world's most widely known optical illusion, where the moon appears considerably larger in the sky when near the horizon. There are records of ancient Babylonians and Greeks debating why the illusion occurs. Some neuroscientists have even begun using MRI studies to understand what's going on.[39]

Or consider how trivially small numbers can be daunting for humans and not for machines. The "Magical Number Seven, Plus or Minus Two: Some Limits on Our Capacity for Processing Information" is one of the most highly cited (more than 33,000 times) papers in psychology. It was published in 1956 in *Psychological Review* by the cognitive psychologist George Miller of Princeton University's Department of Psychology. The paper is often interpreted to argue that the number of objects an average human can hold in working memory is seven, plus or minus two. This is frequently referred to as Miller's law. A simple example is that people cannot quickly memorize random number strings (even if the stakes are high).

The point is that fallibility does not immediately or irrevocably subvert natural intelligence or the commercial viability of machine learning, mainly because nothing is perfect, and nothing can be perfect. Or, to adopt the more poetic language of philosopher Immanuel Kant, "Out of the crooked timber of humanity no straight thing was ever made."

Of course, nothing is imperfect in the same way. That is, various forms of "intelligence" (artificial and natural) are fallible in different ways. For example, in 2015 Google Photos began predicting Black persons as gorillas.[40] It's hard not to catastrophize something so foolish. It's also very easy to overlook intent and measure outcome. However, the principle of charity is a concept in philosophy that encourages people to interpret actions

---

39. S. O. Murray, H. Boyaci, and D. Kersten, "The Representation of Perceived Angular Size in Human Primary Visual Cortex," *Nature Neuroscience* 9 (February 2006): 429–34, doi:10.1038/nn1641.

40. Jessica Guynn, "Google Photos Labeled Black People 'Gorillas,'" *USA Today*, July 1, 2015, www.usatoday.com/story/tech/2015/07/01/google-apologizes-after-photos-identify-black-people-as-gorillas/29567465/.

or statements in the best possible way, whenever possible. To that point, Google did respond quickly, and we must remember that the solutions that don't disappoint are the ones that don't try to solve real-world problems. Google's response matched what was likely their intent.[41]

Google apologized and responded in a *Guardian* article, saying that "image labeling technology is still early and unfortunately, it's nowhere near perfect."[42] The self-evidence of this statement is reflected in the brute-force correction that ignored the root cause of the problem and simply removed the image categories "gorilla," "chimp," "chimpanzee," and "monkey." These terms remain blocked on Google Photos, suggesting that the root cause is not fixed, because it cannot be fixed. We must understand that these solutions aren't racists or chauvinistic. Instead, they are inherently faulty by merely finding relationships between variables and numerous, but imbalanced outcomes.[43] However, spurious relationships between variables and outcomes are not the same as casual relationships between the correct variables and outcomes.

What we need to understand is that machine learning lacks any intelligence or understanding and will therefore fail us, and others, when they leave the lab. The reason is that the real world redefines a problem and places new demands on solutions—demands that will often be difficult to anticipate. Consider that Facebook uses machine learning to rank and recommend news feeds, which creates new problems of ranking and

---

41. Specifically, a Google spokeswoman told the BBC that "We're appalled and genuinely sorry that this happened. We are taking immediate action to prevent this type of result from appearing. There is still clearly a lot of work to do with automatic image labelling, and we're looking at how we can prevent these types of mistakes from happening in the future," https://www.bbc.com/news/technology-33347866.

42. Alex Hern, "Google's Solution to Accidental Algorithmic Racism: Ban Gorillas," *Guardian*, January 2018, www.theguardian.com/technology/2018/jan/12/google-racism -ban-gorilla-black-people.

43. Class imbalance is where classes have significantly different frequencies. For example, a disease data set in which 0.0001 of examples have positive labels and 0.9999 have negative labels is a class-imbalanced problem, whereas a 50/50 split of positive and negative labels is not a class-imbalanced problem.

recommending news that result in a filter bubble and clickbait.[44] However, when solutions fail, we should not hide the failures. We should learn from them. By making human comparisons, we run the risk of black-boxing our understanding and, consequently, we risk even more.

By creating false equivalencies, we outsource responsibility of our solutions to our solutions. During an April 2018 congressional hearing, Facebook's cofounder and CEO, Mark Zuckerberg, referred to "AI" more than thirty times during ten hours of questioning from lawmakers.[45] Zuckerberg said that AI would one day be smart, sophisticated, and eagle-eyed enough to fight against a vast variety of platform-spoiling misbehavior, including fake news, hate speech, discriminatory ads, and terrorist propaganda. However, false equivalencies are not a justification, and solutions are not responsible to us. We are responsible for them.

The world is largely fused, and it has become difficult to maintain the type of neutrality that Zuckerberg seems to promote about solutions. We should certainly wonder about the many entrepreneurs who live by the now famous motto of the Facebook founder, "move fast and break things." It is at least worth considering moving slower and breaking fewer things, which may be optimal over a longer period of time. If Facebook's legal issues over the last few years can teach us anything, it's that we should

---

44. This is why problem-solving strategies are often represented as a cycle (e.g., OODA (observe, orient, decide, and act) loop or PDCA (plan–do–check–act)). The reason is once you have solved a problem another will usually pop up. This is also why Kaggle is a poor representation of real business problems since problem solving happens linearly rather than cyclically.

45. One wonders what a name like "Google Brain" does for accurate evaluations of one's work and the people impacted by that work. Peter Norvig said shortly after the Photo incident: "I think we'd be better off if we had better names other than neural nets and maybe we would be better off if the Google Brain team had a different name. What the Google Brain team provides is programming tools for solving problems—it's not a tool for understanding the brain and it's not necessarily linked to how the brain works," https://www.forbes.com/sites/gilpress/2016/12/21/artificial-intelligence-pioneers-peter-norvig-google/?sh=2099168638c6.

never underestimate the damage created by our solutions, or assume that AI alone will save our planet or provide the answer to misbegotten goals.[46]

Ultimately, asking for forgiveness may not always be better than asking for permission. Though the truth is that as a business manager you may need to ask for both permission and forgiveness. The leadership challenge for all of us is managing the undeniable tension between fallible solutions and getting things done. To find success, we must manage, not merely measure, success. Success is ultimately a balance between starting and finishing and succeeding and failing. There will be aspects of every problem that should not be ignored and other aspects that can only be learned by failing. Therefore, even good metrics are not replacements for good management.

Remember, the human measuring stick is someone else's metric meant to represent their goals. By selecting these odd proxies, we necessarily introduce noise and begin optimization toward strange standards. It tends to be a sloppy metric for businesses because it is a poor proxy for customer development and tends to lead to over-attribution. Human measuring sticks may be useful for insiders as they attempt to understand natural intelligence and ensure accurate self-evaluations. However, business problems need more liberal instrumentation and measurement, while doing as little damage as possible.

---

46. Alex Hern, "Facebook: We Were Too Slow to Recognize Our 'Corrosive' Effect on Democracy," *The Guardian*, January 22, 2018, www.theguardian.com/technology/2018/jan/22/facebook-too-slow-social-media-fake-news-hiring; Celine Herweijer, "8 Ways AI Can Help Save the Planet," *World Economic Forum*, January 24, 2018, www.weforum.org/agenda/2018/01/8-ways-ai-can-help-save-the-planet/.

AI THEATER AND CHILLY WINTERS

It isn't that they cannot find the solution. It
is that they cannot see the problem.
—G. K. Chesterton

Dramatic, *theatrical*, *histrionic*, and *melodramatic* are not words
that should be associated with artificial intelligence, but it does
seem as though some are trying really hard to create a specific
sense of mood and atmosphere when talking about solutions.

Take for example Tech Insider (@techinsider with more than 400,000
Twitter followers). In 2017 they shared a story that they spoke with Sophia,
who's the first-ever robot citizen (now a Saudi citizen) and reported that
"she would destroy humans."

"Oh yeah, she is basically alive," said Hanson Robotics CEO, David Hanson, of Sophia during an appearance on Jimmy Fallon's *Tonight Show*. The adoring and uncritical press that followed those public appearances of Sophia only helped the company grow. Hanson has raised nearly $22 million. Fear it seems is an effective fundraising technique. That is, if you want a well-funded company, all you need to do is create a specific sense of mood and play on the unreasonable fears of maleficent AI, rather than solve a problem.

The truth is that Sophia is a weird technological "sock-puppet."[1] Sophia is a talking head because functioning arms, hands, fingers, and legs are the hardest parts of robotics. Sophia exploits anthropomorphism, even though the ELIZA effect says it doesn't need to.[2]

Based on the inflated claims, how should we evaluate or invest in Hanson Robotics? What does David Hanson mean by "alive" much less "basically alive"? Of course, Hanson Robotics will not answer these questions because they do not want you to know they are full of it. Sentience is not a problem for humans (unless we are dead) and certainly not a problem that businesses have because they are not alive.

AI Theater is a form of customer manipulation akin to so-called vaporware or software that has been advertised but is not available to buy. However, selling hype is not an effective customer acquisition strategy. In fact, selling hype to attract funding, citations, or attention is a type of tragedy of the commons where we are overfishing a pool of possible but finite resources, like attention. Any solution that has to be anthropomorphized is unlikely a solution to anything because it often indicates that the designer cares more about vanity than an actual problem.

In fact, the more extreme the anthropomorphism, the more likely the solution is poppycock. Anthropomorphism plays on our human tendency to project human qualities on inanimate objects. Look at videos of the

---

1. Adam Geitgey, "The Real Scandal of AI: Awful Stock Photos," *Medium*, May 30, 2018, medium.com/@ageitgey/the-real-scandal-of-ai-awful-stock-photos-456633b9b0fc.

2. Even against the unmistakable certitude of being "alive."

machines in the DARPA Robot Challenge.[3] Although these robots cannot open doors, at least they are not pretending to look human or become citizens of nation states. In other words, DARPA cares enough about the problem to not care what their solution looks like.

Of course, AI Theater is easy to accomplish because AI is an umbrella term with a plethora of constructs that conceptually overlap. Consider that artificial intelligence and *cognitive computing*—a term first coined by IBM—is a tautology. In other words, there are no material differences between these differences. Cognitive computing is an example of intellectual dishonesty. In fact, in an effort to avoid labels with negative connotation or adhere to labels with positive connotation in the public eye, many just adopt or change labels, suggesting that there is no endgame aside from hype and manipulation.

There's speculation that IBM chose "cognitive computing" in an effort to avoid the tainted label of "AI" created in part by IBM and oversized claims.[4] As the prevailing winds have shifted, IBM is again using the term *AI*. Incidentally, the 2017 Rittenhouse Rankings of CEO Candor Analytics Survey ranks IBM—a repeat offender—sixth highest in the "F.O.G." or "Fact-deficient, Obfuscating, Generalities" category.

Today, there are strong incentives because AI start-ups receive 15 and 50 percent more funding compared with other companies, and entrepreneurs race to complete robust exits that look more like financial arbitrage than building businesses and solving problems.[5] For example, how do we evaluate a company like Darktrace, a cybersecurity company valued at $1.65 billion that claims to be modeled on the human immune system?

---

3. If there was a robot revolt in the near future, just be sure to close the door behind you. *See* Erico Guizzo, "A Compilation of Robots Falling Down at the DARPA Robotics Challenge," YouTube, June 6, 2015, www.youtube.com/watch?v=g0TaYhjpOfo2.

4. R. Korf, "Does Deep Blue Use AI?" *AAAIWS'97–04*, in Proceedings on the 4th AAAI Conference on Deep Blue Versus Kasparov: The Significance for Artificial Intelligence, January 1997.

5. A business exit strategy is an entrepreneur's strategic plan to sell his or her ownership in a company to investors or another company.

The human immune system is amazingly complex and has evolved over millions of years. It can recognize and remember millions of different enemies, and it can produce secretions and cells to match up with and wipe out each one of them. Darktrace certainly lacks this degree of coverage (i.e., the ability to recognize and remember millions of different enemies) and lacks the "offensive" capacity to match up with and wipe out each enemy.

AI Theater is a race to the bottom for the most disingenuous. However, attention is scarcer than ever, and the race for cheap attention is a race that can't be won. As soon as someone gains an advantage, someone else will lower their standards to obtain more attention and bad will push out good. Gresham's law is a monetary concept that highlights how bad or unrealistic claims ("bad money") force out good and realistic claims ("good money").

However, these misfits have already lowered their standards and self-esteem, so it's an escalating hijacking of trust. The end result is AI Theater, where at least some choose to avoid solving problems at all cost by overselling their solutions in dramatic ways. Once-respected media outlets now promote shameless headlines, and entire industries are based on clickbait, false and empty claims, and vaporware.

Theater is not reserved for the disingenuous or those trying to deceive; it applies to the dim or self-deluded as well. In a 1976 paper titled, "Artificial Intelligence Meets Natural Stupidity," Yale University professor Drew McDermott coined the term "wishful mnemonic" to refer to the simple-mindedness of describing solutions by their grandiose pursuit.[6] Consider open-source software with labels like "OpenCog" (short for open-cognition), which can mislead others as well as themselves.[7] Although OpenCog describes itself as a "framework for integrated artificial intelligence and artificial general intelligence," to a great degree such programs

---

6. Drew McDermott, "Artificial Intelligence Meets Natural Stupidity," *ACM SIGART Bulletin* 57 (1976): 4–9, doi:10.1145/1045339.1045340.

7. OpenCog is an open-source project launched in 2008 by Ben Goertzel and Cassio Pennachin. Recall from earlier that Goertzel and Pennachin edited a book titled *Artificial General Intelligence*, which called for a resurgence to adhere to the original vision of AI.

are not solutions as much as they are attempts at describing a problem in a new way.[8] In this case, the problem of cognition. In other words, despite what the project wants us to think, OpenCog does not represent a better way to implement some well-understood task like cognition. Rather, Open-Cog is an attempt to program the first implementation of such a solution. And because no solution possesses general intelligence, no framework for integration is required.

Whether we're aware or unaware, accidentally or inadvertently misleading ourselves or others and damaging our business or harming customers is inexcusable. For example, mixed and incompatible metaphors perpetuated by misleading stock photos are a sure sign that someone is not interested in what he or she is saying. Adam Geitgey, author of *Machine Learning Is Fun!*, focuses on the absolutely atrocious stock photos used in most news stories and many PowerPoint presentations on AI.

We are all likely familiar with photos that anthropomorphize solutions (and, perhaps, guilty of using them ourselves). Examples include the human brain transposed with computer circuitry (incorrectly suggesting one type of intelligence with solutions and one type of solution using neurobiological computing); a human reaching out to a robot akin to God providing a spark of life from His own finger to Adam's, as in Michelangelo's painting on the ceiling of the Sistine Chapel; or, a personal favorite, the incoherent nature of a robot using a computer.[9]

Modern metaphors that compare solutions with carbon-based organisms without any additional evidence should always raise a red flag. We can't solve problems with metaphors. Metaphors are not evidence, and they lead to oversimplification. For example, metaphors that describe machine learning in ways similar to human learning are misplaced. Machine learning is very brittle, and it requires a lot of preparation by data scientists,

---

8. https://github.com/aSchimp/opencog#readme.

9. How much good did Big Data metaphors like "oil drilling," "gold mining," and "dragnet fishing" do for us all? Big Data metaphors like "mining" often, and incorrectly, presuppose interpretation of data, fault-tolerance, and unassailable objectivity.

special-purpose coding, special-purpose data sets, objective functions to optimize, and custom learning structure for each new problem. All of this discounts the harder aspects of deploying, implementing, and scaling any solution in the enterprise or for commercialization. Even after all of this work, we are left with only statistical approximations that interpolate signals that never represent intelligence.

Ultimately, I get it. We all want external audiences to view our work as easy and magical. Most investors and many early adopters want to be wowed by the perceived ease of problem solving. But this perception does not result in direct advantages or direct benefits because the early (and late) majority of customers will care more about "does it work" than "is it new." Although naming a solution "AI" is in vogue and much easier than problem solving, the ease of talking only highlights indirect advantages. That is, talking may strike an important chord with customers who use emotion in their decisions, but we ought to recognize the origin of indirect advantages versus the direct advantages of the explicit design and problem solving.

The philosopher Friedrich Nietzsche wrote, "With everything perfect we do not ask how it came to be." Instead, "we rejoice in the present fact as though it came out of the ground by magic." Nietzsche observed that our vanity promotes a culture of genius and wrote that "if we think of genius as something magical, we are not obliged to compare ourselves and find ourselves lacking."

Ultimately, there is a mystique that surrounds seemingly effortless, magical performances. We marvel at how some solution is able to do things that we cannot easily do. We create hero-solutions and ignore the amount of time it took designers to continuously problem solve and management's role to create the required cultures, and the overwhelming commitment by everyone required for execution at scale. When we examine end results, we interpret the story backward, starting with the solution and ignoring the work of problem solving. Managers interested in creating direct advantages will need to dispel magic and create hero-problems rather than hero-solutions.

It's Time for a New Golden Rule

The term *AI winter* refers to the chilly periods of disillusionment and disinterest in AI that follow periods of massive disparity between the expectations and realizations of AI promises. Just like an asset bubble, the value of the AI industry as a whole pops as people collectively realize that our solutions, while not exactly worthless, are worth significantly less than they thought; regardless if there is still value. Previous AI winters resulted from hype and rapid investment, when over time bad money and pomposity drove out good money, modesty, and real problems. Even if you have not lived through an AI winter, you can still learn from it.

AI Theater produces the wrong kind of emotion and energy that inevitably leads to inaction, a type of solution sloth. That is, amid the euphoria to satisfy ambitious external goals, a business's purpose gets lost. We give attention elsewhere even if it was us that failed to allocate enough time to noticing problems. We wise up to the unfulfilled expectations and turn overly cautious. In other words, the increase in uncertainty does not lead to an increase in the probability that we will fail with our initiatives but that we will fail to even launch initiatives to solve some of our most important problems.

Consequently, I'm advocating for an AI golden rule as a way to moderate and protect against overly optimistic dreams and fear-based manipulations, both of which lead to market boom-bust cycles in AI. The AI golden rule is to talk about a solution in the open, candid way we'd want solutions to be discussed with us. It is a strategy of selling a solution based on a real value proposition, rather than on fear or on the external dream of a solution.

Said differently, if you want customers to listen, you need to communicate through mechanisms they already understand, like their problems, and communicate clearly the value proposition. In fact, if you can describe your customers' problem better than they can, they will assume that you know how to solve that problem. In other words, we don't need to talk about exaggerated solutions using incoherent metaphors that mislead, because as Steve Jobs said, "If you define the problem correctly you will almost have the solution." You will also have much of the customer acquisition strategy.

Of course, the golden rule is not a panacea. It rests upon the basic assumption that other people would like to be treated the way that we would like to be treated. Perhaps Hanson Robotics, IBM, Tech Insider, and OpenCog assume we like being misled, because they themselves like to be misled? After all, those who mislead others in public will eventually mislead themselves in private. Nonetheless, the AI golden rule is a chance to work toward the top rather than the bottom: to deliver personal and timely messages to customers that matter.

PART
TWO

PROBLEMS AND
PROBLEM SOLVING
WITH AI

NOT ALL PROBLEMS ARE CREATED EQUALLY

Problems are not stop signs, they are guidelines.
—Robert H. Shuller

The final section aims to serve as a guide for understanding problems and provide approaches to problem solving. What I have found is that making sense of AI requires more than immersion in the technical details of the newest solution. It requires a sense of how problems impact our solutions, how problems can differ, and how to approach problem solving.

The main reason we will focus on problems is simple. The most successful managers—and by extension, successful businesses—find the right problems to solve, and solve them with the right solution and the correct amount. Although it is tempting to be solution-focused, especially when it comes to AI, being at the leading edge of an industry doesn't mean you have to be—or even should be—one of the small number of insiders pushing solutions forward. You can be at the leading edge as a user or the leading edge of a problem.

So, what then is a problem?

According to the Oxford English Dictionary, problems are defined as "a doubtful or difficult matter requiring a solution" and "something hard to understand or accomplish or deal with."

In other words, problems are undesirable situations that occur when something is wrong. For business this will often mean something undesirable to customers or potential customers. At times these undesirable situations will include internal problems, which you will invariably have greater intimacy with because you are the customer. However, if you wish to move a solution for an internal problem to an external audience, you should always ask whether your problem is one an external audience has and can solve themselves. In other words, internal problems require you to determine whether you and your problem are representative of some pervasive problem experienced by others. It may not change the fact that some of the best problems are those that you have already had the fortunate misfortune of living with firsthand.

Although the Oxford definition suggests that problems are only problems when the solution is not obvious ("something hard to understand or accomplish or deal with"), this isn't always the case because even well-defined problems with obvious solutions may still need solving. After all, even obvious, easy problems do not solve themselves. Of course, we should not assume that what is obvious to us is obvious to everyone, but the assertion does make intuitive sense from a business perspective. That is, an obvious solution may be obvious to everyone including potential customers. In other words, not all problems are good business problems because not all problems create customers. Therefore, problems may include those well-defined problems with obvious solutions—which may lack an external audience but may increase efficiencies or lower cost—and business problems that are ultimately market-relevant problems with nonobvious solutions.

Market-relevant problems often result from situations that others don't want to exist, but can't make go away, because the problem is something hard to understand and deal with. In other words, problems you may want to monetize aren't just problems; they are problems that others want solved but are unable to solve—or are uninterested in solving—themselves. Business problems also include problems customers want solved and are solving but not as well as they'd like. As noted earlier, problems worth solving are sometimes created as by-products of old solutions as a type of innovator's gift.

Solutions that can be monetized are not always obvious solutions, because someone has either already solved the problem or decided to live with it. The two types of paying customers are those who will live with the consequences of a solution—or a partial solution—and those who will live with the consequences of no solution, not for lack of interest in solving the problem but because they can't solve it.

The biggest problem with problems is that we often see them as one thing, just as the Oxford English Dictionary considers problems abstractly as just one thing: a problem. The problem with the single-problem mindset is that we also think of solutions as being one thing and consequently view problem solving as a fixed process. However, there are all types of problems with various characteristics and degrees of solvability.

1) Consider that not all problems are even problems. Recall that impostor or fake problems are sometimes created with solution-centric thinking. However, real problem solving does not create problems to solve; it is always a response to a real problem that already exists. Needless to say, impostor problems are not good business problems because they are not problems at all. The most common mistake with solution-focused strategies is guessing at a problem to solve that no one actually has with solutions we find important. In other words, we

find some solution so important that we create problems for them to solve. However, problem guessing is not an effective problem-solving strategy. Just because you can solve it, doesn't mean anyone will care.

2) The costs of creating problems to solve, even with interesting solutions, is the opportunity cost and the creation of orphan problems. Orphan problems are the problems we ignore while we work on fake problems.

3) Closely related to impostor problems are Goldilocks problems. Goldilocks problems are neither too easy nor too hard but "just right" for our solutions. They are not entirely fake problems, but they are not exactly real, either. In *Goldilocks and the Three Bears*, author Robert Southey shows that Goldilocks solves her problems of being hungry and tired by always making the "middle" choice. To solve her tiredness, she tries out each solution and determines that one bed is too hard, one is too soft, and one's just right. In other words, Goldilocks selects the optimal solution after sampling the extreme solutions for her problem. Of course, our solutions should always be *just right* for our problem: a type of problem-solution fit. However, when this process goes backward, misalignments are created. Imagine a solution in search of the perfect problem, determined after sampling from the extremes that are neither too hard nor too easy to solve. In this way, solutions will be manufactured in the sense that we are working backward from a solution to the perfect problem, which gives us only a poor replica of the real world. If some arbitrary solution shifts our focus away from a real problem and places an emphasis on the perfect problem, then our strategy is solution-focused—which rarely supports goal alignment. And it's also a Goldilocks problem. Consider how games, contests, and benchmark data sets under sample real-world problems, although real, are neither too hot, too cold, too small, too big, too hard, or too soft, but ultimately just right for a solution. The real world does not always (or even often) produce Goldilocks

conditions, and businesses do not need to work backward from solutions to the right problem for those solutions.

4) There are "real" problems. Real-world problem solving will return you to and inform your understanding of the real world while advancing your understanding of a real-world customer.

5) There are problems for which the problem itself is obvious. These are *easy problems* because they are intrinsically well-defined problems. Our goal is always to define our problems; however, intrinsically well-defined problems may not always be good business problems, because easy problems can be easily seen and often easily solved by others. In other words, those impacted by an easy problem may be poor customers as they may not be motivated to buy something they can see and perhaps do themselves. Ultimately, not all problems that can be solved, should be solved, because not all problems have market relevance.

6) There are *hard problems*, which are ill-defined problems. Hard problems cannot always be defined in a complete manner. Ill-defined problems often result in clumsy solutions, but because the problem is difficult to understand and define, some solutions find market acceptance.[1] Market acceptance is achieved because customers are unwilling—and often unable—to solve a problem themselves.

7) To be sure, there are problems with our solutions, too, as solutions are never perfect. However, the specific desire to possess better-than-needed solutions is solution solving. Solution solving is largely the domain of insiders, who create external goals for solutions that are often disconnected from any specific problem. Problem solving our

---

1. Marco Verweij and Michael Thompson explore the nature of solutions—what they call "clumsy solutions" within complex problems: M. Verweij and M. Thompson, eds., *Clumsy Solutions for A Complex World* (New York: Palgrave Macmillan, 2006), doi:10.1057/9780230624887.

solutions without a problem produces strange (strange in the sense that they may be too large, complex, or difficult to implement; or impossible to debug, interpret, or explain) solutions that may result in no real-world application, because a problem can't be found—partly because no one was ever looking for a problem in the first place. Remember that creativity is not the best goal, commercialization is, which requires the creation of value for a customer who rarely, if ever, cares how that value is created.

Moreover, when a solution is disconnected from a problem, it is difficult to evaluate improvements. A solution should have only what is needed to solve a problem, not everything! The goal of making a solution better than it needs to be in order to solve a bigger problem—or more than one problem—often reflects that we haven't acquired a problem to begin with. Instead, we make the problem as big as possible in hopes of acquiring a customer by solving anything. The thinking goes, "there has to be something that someone will find interesting," but customers often detect the inauthenticity of trying to be all things. For businesses, problem solving with solutions begins by asking why a customer cares, not merely justifying work by making a solution more complex than it needs to be.

8) There are also simple problems and complex problems. We will discuss the nature of these problems in depth. However, we may ultimately consider *simple* and *complex* problems "ordinary" when compared to *wicked* problems. Wicked problems have chaotic properties, cannot often be solved, and cannot be broken down into smaller parts. We will also discuss the nature of wicked problems in more detail.

Ultimately, if we lack problems, we are no longer doing good business. If we lose sight of a problem then we have lost sight of our goal, and if we lose sight of the person impacted by a problem, then we will also lose sight of a customer. The reason to discuss various problems rather than just one kind of problem is that from a management perspective ignoring various problems is shortsighted. It is shortsighted in the sense that we will not generally know where our next business model or customers will come from. The most effective problem solvers will avoid the trap of a single problem.

Real Problems and Real Solutions

Noticing problems is the ability to see what is wrong, what is missing, and what is potentially unnecessary to add. Of course, this requires us to dedicate time, which means noticing problems is also a real problem. Decades of evidence shows that managers are time starved.[2] Although time management is a staple of efficient business, it is often toxic to a manager's ability to notice and contemplate problems. In other words, being busy and assuming multiple roles are essential to managerial work, but to be effective at noticing real problems it is also critical to allocate time, attention, and resources.

How much time we spend on noticing problems depends on how much we already know and want to know about a problem. The optimal managerial time management for noticing problems ultimately depends on the time-fidelity trade-off you're willing to make.[3] If the problem is well-defined or one you already know well, then you don't need to spend much time understanding it. There is already a great deal of prior knowledge and thus certainty. However, if you lack firsthand knowledge of the problem, or if it is ill-defined or complex, then you need to acquire certainty about the

2. O. Bandiera et al., "A Survey of How 1,000 CEOs Spend Their Day Reveals What Makes Leaders Successful," October 12, 2017, *Harvard Business Review*, https://hbr.org/2017/10/a-survey-of-how-1000-ceos-spend-their-day-reveals-what-makes-leaders-successful; H. Bruch and S. Ghoshal, "Beware the Busy Manager," *Harvard Business Review* 80, no. 2 (February 2002): 62–69; H. Mintzberg, "The Manager's Job: Folklore and Fact," *Harvard Business Review* 53, no. 4 (1975): 49–61.

3. Juergen Sauer and Andreas Sonderegger, "The Influence of Prototype Fidelity and Aesthetics of Design in Usability Tests: Effects on User Behaviour, Subjective Evaluation and Emotion," *Applied Ergonomics* 40, no. 4 (2009): 670–77, https://doi.org/10.1016/j.apergo.2008.06.006.

problem. This becomes a trade-off between how much you need to know and how much time you have to allocate to the acquisition of certainty about the problem.

The more time you spend with a problem, the more comprehension about that problem you will acquire. The amount of detail gathered will invariably guide what solution will be produced, and therefore we ought to approach this essential component with pragmatism. Ultimately, there is no magic number, and this trade-off depends entirely on resources available, the certainty required, and the difficulty of the problem. Neither extremes of "paralysis by analysis" or "extinction by instinct" are desirable.[4] The former can exact a high opportunity cost from over-analysis while the latter risks a poor or incomplete response. This process is mostly about finding some middle ground between spending too much time and not enough. Viewed this way, solution guessing—and problem guessing—is low-fidelity design in that we do not spend enough time on the problem.

Although having certainty, firsthand experience, or prior knowledge of a problem is evidence of a problem's authenticity, it does not guarantee that anyone else has the same problem. Your problems may be real to you, and you deserve to have them solved, but that doesn't necessarily mean that others share them. Although solution-focused thinking often fools us into thinking that fake problems are real because they can be solved, real problems sometimes fool us into thinking that we are not alone. Said differently, you are not a market and finding a solution to a real problem experienced firsthand is not market relevance. You must still figure out if your problem is something shared by another if you seek market relevance.

Central to a real problem is that the problem exists in the real world. We are told we should care about someone else's solution that exists in a

---

4. A. Langley, "Between 'Paralysis by Analysis' and 'Extinction by Instinct,'" *Sloan Management Review* 36, no. 3 (1995): 63–76.

lab or feeds on academic benchmark data sets. Often, we shouldn't. Problem solving seldom functions in these environments because it often fails to take us back to a specific problem. Try asking yourself if the problem is real. Ask yourself if anyone else has this same problem. By answering these basic questions, you will be able to answer whether the problem is real and whether it needs to be solved.

Furthermore, real problems produce solutions that are much sloppier than we typically think. In fact, although real-world solutions exist in the real world, they are paradoxically much more difficult to identify. They are notoriously difficult to discern, analyze, and understand, and they are much harder to imagine before we solve a problem. Consider a solution that is entirely rhetorical: the fictitious solution used for a food app in the popular HBO show *Silicon Valley*. The characters Erlich and Jian-Yang of the fictional SeeFood start-up create a solution for the contrived problem of identifying hot dogs. The solution was to become the "Shazam for food," but ultimately it just told users if the object they pointed their phone's camera at was a hot dog or not-hot-dog. To be sure, all solutions have humble beginnings, and this is a humble starting point.

The problem of predicting hot dogs is an easy problem because it is a well-defined problem. It's not hard to imagine the solution to this so-called problem as being a single algorithm. Moreover, the solution can likely be trained in a single episode of learning and may never get any feedback on performance, primarily because it is unlikely to need it. The problem is simple in that it contains no arbitrariness between inputs and outputs. That is, the features that predict a hot dog are static over time: a hot dog is always a hot dog, and the features that predict a hot dog are always the features that predict hot dogs and not salads.

Furthermore, misclassifying a boot as a hot dog is unlikely to result in significant issues for users (unless those users are silly enough to eat the boot), so no one needs to take responsibility for the safety, security, interpretability, or explainability of the algorithm. There would be no reason, for example, to call Homeland Security if there was a data breach or adversarial hot dogs were introduced into the data pipeline or at run-time. The point is that when the problem is fake, or at least trivial, the solution will be much more tangible and much easier to understand than a solution to more complex, real-world problems.

Contrast the Shazam for food with Facebook's News Feed. News Feed highlights information for users among all information, thus positively impacting so-called information overload. Facebook's proprietary solution compares the merits of about two thousand potential posts every time Facebook is opened, using a complex system meant to provide a meaningful experience over that of clicks, reactions, or reading time. News Feed has also been described as a filter bubble, showing users personalized results about information deemed interesting, in contrast to showing all information, even information users may disagree with.

Consequently, Facebook has been criticized as effectively becoming the world's largest echo chamber, which has a whole host of social implications given Facebook's scale.[5] Facebook has been researching the issue of a filter bubble since 2010.[6] By late 2013, clickbait articles had become significantly prevalent, forcing Facebook to add a number of data points to its algorithm to significantly reduce them. Facebook has also hired thousands of employees to assess and monitor News Feed, because the context of this problem—specifically the social context—is so important. Context is important because real-world solutions impact how we work and how we live. In all, Facebook has to resolve the paradox of managing a variety of pursuits like information overload and providing users information they

---

5. Megan Knight, "Explainer: How Facebook Has Become the World's Largest Echo Chamber," *The Conversation*, May 18, 2020, theconversation.com/explainer-how-facebook-has-become-the-worlds-largest-echo-chamber-91024. Jean Twenge, professor of psychology at San Diego State University, notes how social-networking sites like Facebook promise to connect us to friends, but many teens are emerging as a lonely, dislocated generation: *iGen: Why Today's Super-Connected Kids Are Growing Up Less Rebellious* (New York: Atria Books, 2017).

6. Farhad Manjoo, "Can Facebook Fix Its Own Worst Bug?" *The New York Times*, April 25, 2017, www.nytimes.com/2017/04/25/magazine/can-facebook-fix-its-own-worst-bug.html.

seek, while also avoiding echo chambers by ensuring information diversity and serendipity.

Silicon Valley's Shazam for food provides no ontological, economic, or social value. It's a gimmick. The solution does not learn from its mistakes because it doesn't have to. Shazam for food never requires a reframing of the problem, because the problem is fake. It will never see new classes and new data, or have any significant impact on our lives. Moreover, because the context of problems often matter, we can assume that Shazam for food does not. That is, the context of what is and what is not a hot dog does not matter, because the "problem" is not a real-world problem that anyone would identify with today.[7] We routinely distinguish hot dogs from hamburgers and after learning what a hot dog is, we will never again not know.[8]

The point is that real problems create sloppy, clumsy solutions. Real problems end up redefining both the boundaries and the essence of the original solution because implementation actually confronts the real world. For many real problems with even moderate complexity, what this all means is that the data we have is not often the data we need. Therefore, we often have to reframe our problem to find a problem that we can support with data. Then upon training we may realize that we need more data or need to relabel it, and when we deploy our solution, we often realize that we need to add more classes or other supportive elements. If you are not wrestling with the framing and reframing of a problem, the data you have versus the data you need, missing classes, or how a solution performs when deployed, then you may not solve a real problem.

---

7. Paul Graham calls impostor problems "made-up" or "sitcom" startup ideas. *See* "How to Get Startup Ideas," http://www.paulgraham.com/startupideas.html#f10n.

8. Clayton Christensen in *Competing Against Luck* highlights the importance of "progress" as movement toward a goal. For Christensen, a "job" always demands progress and is rarely a discrete event like the Shazam of food. That is, even if you don't know what a hot dog is, you will never again not know. The problem does not repeat itself and therefore customers will not require progress to be made. If the problem doesn't require progress, then it is not a good business problem according to Christensen. Clayton M. Christensen et al., *Competing Against Luck* (New York: Harper Business, 2016), 34, Kindle edition.

Consider the Facebook example again: its problem became another problem that then became a paradoxical problem of managing information overload and providing users information they seek while avoiding echo chambers. This is not to excuse Facebook and their troubles managing the problem, but it is meant to highlight that real solutions often break down at scale and create a host of new problems that require modification to the original solution. In fact, if some solution does not change when deployed, it is probably not solving a real problem. Or, if the problem is real, but you don't feel like you need to change your solution, then the problem may lack importance. The desire to "get it right" is a key characteristic of a real problem that may create a customer. If you don't care enough that the solution is right, no customer is going to care about the solution.

The revelation upon deployment that we knew much less about a problem than originally thought is a hallmark trait of a real problem. In fact, if some solution is exactly the same in the lab (i.e., no difference between training, test, and deployment) as you'd expect to find in the real world, then you may be working on an impostor problem. The Shazam for food does not require any redefinition of the problem, because it's not a real problem. The only way to understand how a solution works in the long term is to have it train on the data it acquired when it was in the wild. This is very difficult to do, and performance varies greatly.[9] Facebook has to solve paradoxes because it is solving a real problem and performance broke down over time.

All of this is to say that if you're not working on a real problem by creating a real solution, then you're distracted by the means and not working on the ends. When we work on working or practice for the sake of practicing, we quickly hit ceilings, because there is not a real user that needs us to solve a real problem. Many organizations may already have "experimentation" environments or "innovation labs" that create good atmospheres for

9. Martin Zinkevich, "Rules of Machine Learning: | ML Universal Guides | Google Developers," Google, June 12, 2019, developers.google.com/machine-learning/guides /rules-of-ml/.

solutions, but we must always remember what they lack. They lack a real problem and a real user.

Even when we put a real problem and user into these environments, it does not ensure that either will respond the same way they would in a real-world environment with real constraints. We will likely get a different result in the laboratory if we attempt to take a squirrel from its natural habitat and study it. Ecologies are so intertwined that it is impossible to isolate any component without affecting the analysis. Many problems are the same way.

In other words, do not make problem solving easier by making it fake. If you need practice, find a real problem. Otherwise the lessons you learn will be synthetic, just like the so-called problem. Impostor problems will not challenge your team to correctly frame a problem because without a user there can be no user development. As Einstein apocryphally said, everything should be made as simple as possible, but not simpler. By making problems easier to the point of making them fake, we reduce the basic parts of a problem to something meaningless and lose something of vital importance. Do not get trapped in pilot purgatory, an endless cycle of practicing on fake problems.

Is the Problem Simple?

When it comes to easy problems, "easy" is not meant pejoratively as inconsequential. Nonetheless, easy problems are *easy* because they are approachable and inherently well-defined. Our goal should always be to define our problem, and easy problems make this process much more amenable.

In other words, an easy problem is a well-defined problem in which the initial state, or starting point, and the final state, or goal, are clearly discernible within the problem itself.[10] When the problem is well defined,

---

10. While *easy* may often be interpreted as simple, simple is the opposite of complex. So-called complex problems include those problems that are varied, multiple, and interwoven together with other problems with various goals. "Simple" suggests that the problem is without much ambiguity and thus any subsequent solution will not optimize conflicting goals.

a description of a problem can be easily created. Ignore for a moment that the problem solved with the Shazam for food is an imposter problem and consider how it is a well-defined problem. The task of predicting a hot dog is a problem that includes the goal. The problem of predicting hot dogs has already been half-solved in that it has already been defined. As former head of research at General Motors Charles Kettering said, a problem well stated is a problem half-solved. It may sound odd to have a problem that is half-solved, but it is true if we consider that the problem has no ambiguity and contains complete information.

Keep in mind, however, that the world of half-solved problems is not only stripped down but also commoditized. That is, intrinsically well-defined and familiar problems are being commoditized by Google, Microsoft, and Amazon (to name a few), which are making machine learning and computing more accessible to customers willing to solve simple problems themselves. To be sure, machine learning makes this half much easier to solve, but monetization requires someone incapable and unwilling to comprehend problem themselves.

Consider that Kewpie, a Japanese food manufacturing company, trained an artificial neural network to identify defective potatoes. Kewpie uses a video feed from existing production lines to monitor potatoes. The solution is trained using data that are labeled images of defective and quality potatoes. The solution identifies defective potatoes with a similar accuracy to human quality-control experts at a scale comparable to humans. The project ultimately resulted in savings of more than $100,000 per production line and took just six months to implement.

Kewpie's problem is easy, as it is well-defined (i.e., detect defective potatoes). Kewpie's problem is also simple. A simple problem is a problem that cannot be decomposed into smaller pieces. Consequently, simple problems often have the easiest implementation and return-on-investment (ROI) calculation. In fact, we can probably assume that any problem in which one object moves at a slow fixed rate in one direction on a conveyor belt with no changes in context is a simple problem. Too bad that few problems exist on conveyor belts.

Ultimately, this type of closed, ordered system is vastly different from all the things happening on your enterprise network, your local hospital, or every time you drive to work. Often the context in which our problems live can tell us a lot about how hard they will be to solve. Kewpie's problem is a very common problem in computer vision. However, although Kewpie's problem is computer vision and simple, not all computer vision is simple.

We may wonder if Kewpie's solution is AI. Recall, "AI" is an attribution made after the fact based on explanations of why it is intelligent. More importantly, Kewpie is solving a real problem at scale that aligns with their business and aligns their business goals with performance that impacting their bottom line. We know that the problem is not an impostor because it's saving $100,000 per production line. The solution may lack intelligence, and may not be as important, complex, or as significant as Facebook's News Feed or Google Search. If these characteristics are absent in Kewpie's solution, it is more a consequence of the problem and the environment of that problem rather than the solution lacking some grandiose, external goal for itself.

Should we care if Kewpie wanted to call this solution "AI"? The most important aspect of any solution is not what we name it but that it solves a problem and impacts the business as measured against where we started. Recall that a solution is only a solution after it has been evaluated positively against a problem, and problems are only real problems when they have been evaluated positively by the business and align with a business's goals. Ultimately, what Kewpie wants to call their solution is less important than the impact to its business, and they have impacted their business. We must conclude that the problem of detecting defective potatoes is a real-world problem with real economic value.

Artificial intelligence (specifically machine learning) has made an easy problem like detecting defective potatoes much easier to solve over traditional programming, GOFAI, and even traditional statistical modeling. In other words, although the problem of detecting defective potatoes is intrinsically well-defined, the noncoding work is made even easier using machine learning, because we don't have to explain with painstaking detail what constitutes a defective potato. The goal of machine learning—specifically supervised machine learning—is a simple one, using features extracted from good and bad potatoes and labeled data to predict each category evaluated, in this case, using predictive accuracy.

Kolmogorov complexity is a measure that states that the smallest description needed to represent data is often the least complex. Kolmogorov complexity describes a problem by the size of the smallest solution that is able to describe the relationships between the input data and the output.[11] In the worst case (i.e., most complex), it would be necessary to list all the characteristics of the data. However, if there is some regularity in the data due to some obvious structure to the problem, then a compact representation may be obtained.

In other words, machine learning can extract useful representations, which means machine learning solutions are sometimes (not always) less complex than rules for which we might have to list all the characteristics of the data.[12] Kolmogorov complexity ultimately provides a useful heuristic for optimal supervised machine learning problems. That is, if writing down rules to categorize something like a defective potato is difficult, but

---

11. Li Ming and Paul Vitanyi, *An Introduction to Kolmogorov Complexity and Its Applications* (New York: Springer, 1993), 38; Li Ling and Yaser S. Abu-Mostafa, "Data Complexity in Machine Learning" (technical report, California Institute of Technology), CSTR: 2006.004.

12. Although in practice Kolmogorov complexity is incomputable, meta-learning research has developed some approximations based on the computation of indicators and geometric descriptors. *See* Tin K. Ho and Mitra Basu, "Complexity Measures of Supervised Classification Problems," *IEEE Transactions on Pattern Analysis and Machine Intelligence* 24, no. 3 (2002): 289–300. *See also* Sameer Singh, "Multiresolution Estimates of Classification Complexity," *IEEE Transactions on Pattern Analysis and Machine Intelligence* 25, no. 12 (2003): 1534–39.

gathering examples of defective potatoes is not, then we likely have a good-use case for supervised machine learning.[13]

Imagine how difficult it would be to design a solution with the ability to solve what is an otherwise ineffable task (e.g., predicting defective potatoes). After all, what is a defective potato? Imagine the difficultly of codifying all aspects of knowledge of a defective crop, such as late blight, early blight, black scurf, dry rots, powdery scabs, and charcoal rots. You would have to learn far more about the problem than you would need to know intuitively to solve the problem. You could probably identify a defective potato, but you might not be able to describe a defective potato. Although this problem may require lots of prior knowledge to design a solution—which may not be trivial to acquire—supervised machine learning can learn the task via labeled data and learning (i.e., minimizing error) good and bad crops.

A more prescriptive view of simple problems is that they are simple due to their scope (i.e., boundedness or closed domain), directness (i.e., uni-task), lack of ambiguities of an output (i.e., universality and certainty), and consistency of inputs over time. Simple problems will often be reflected in easier-to-acquire performance gains and more consistent performances over time. High performance is achievable because the problem is discrete, bounded, often static, and narrow. Because it is hard to solve a problem that a human hasn't already solved, simple problems can succinctly and intuitively be thought of in the following manner:

1) One person can quickly and easily complete the task.
2) A group of people can complete the task faster (i.e., horizontal scaling).[14]
3) That group can universally agree on a solution set for the task.

---

13. Daryl Weir, "Fantastic Problems and Where to Find Them," *SlideShare*, July 3, 2017, www.slideshare.net/futurice/fantastic-problems-and-where-to-find-them-daryl-weir.

14. Horizontal scaling means that we can scale by adding more people to our pool of resources.

Said differently, one person can solve a simple problem because the size of the problem is complete. There is no need to cleave the problem into smaller pieces because we are already working on the smallest batch of the problem, which also happens to be the complete batch of the problem. A defective potato is the smallest part of detecting defective potatoes. When the smallest part of the problem is the complete problem, knowledge can be easily shared. Moreover, when anyone can perform well at a problem, knowledge can be acquired easily. When knowledge can be easily shared and validated, universally agreed on, vertical scaling—that is, a problem that requires adding more qualified personnel—is not required.

When vertical scaling is not required, humans are interchangeable due to the easy access to knowledge of the problem and the degree of universality specific to that problem. Interchangeability means that the problem can be parallelized without any loss of performance and scales horizontally by adding more people to a pool of resources. When the problem is distributed the solution can be parallelized. That is, throwing more workers at the problem will scale linearly.

Alas, complete automation is often possible for simple problems. However, complete automation is likely the first thing that will come to mind for many of us. If your goal is commercialization, try to discount the first thing that comes to mind, especially when that first thing is automation that replaces human workers. You can always come back to it, but not if your resources are occupied by simple problems.

In fact, consider discounting the second, third, and fourth ideas, too, especially when they involve replacing flesh-and-blood people. These ideas are all likely low-hanging fruit. Low-hanging fruit is easy pickings for everyone. After all, when the minimum acceptable performance is low for you, it is low for everyone. Problems that can be solved with near perfect performance with little amount of training data with few dimensions to track and few possible variations in outcome will be easy for anyone to solve. Ultimately, each idea that you discount in pursuit of other, more complex problems, will increase your uniqueness and your competitiveness. You should always stretch to increase the scope of what you're solving. The most important problems for you to solve may be hard, ill-defined problems. Over time your ability to define harder and more complex problems will improve because you will become the type of person who solves complex problems.

Is the Problem Complex?

Hard problems are ill-defined problems. Ill-defined problems are hard because of the nature of the problem, and acquisition of knowledge relating to that problem is difficult to acquire even for those who have the problem. In other words, hard problems are complex problems because the complexity intrinsic to the problem challenges acquisition, framing, and definition.

Consequently, hard problems lead to exclusivity. Because fewer people understand hard problems, few will try to solve them despite potentially higher competitive advantages in the market from possible solutions. Parkinson's law of triviality explains that people will disproportionately gravitate to easy problems, thereby allocating more attention, time, and effort to problem solving.

The law of triviality describes that the amount of talking is inversely proportional to the difficultly of a problem. Or, the harder the problem, the more people shut up. This is exactly what you do not want to do. Parkinson provides an example of a fictional committee whose job it is to approve plans for a nuclear power plant. Instead of discussing nuclear fission they spend most of their time determining what building material to use for the staff bike shed.

Easy problems are accessible, intrinsically well-defined and already half-solved, and their solutions can easily be evaluated in terms of solved and not solved. Therefore, if you find yourself or your team spending a lot of time during a meeting discussing a solution to an easy problem, you are likely wasting resources. That's because your problem is accessible, and if your problem is accessible, talking about doing is less valuable than doing. When something is easy and simple, do not add to complexity, second-guess yourself, or procrastinate. Start solving. Conversely, hard problems are ill-defined, and these problems are not half-solved until we do the work in acquiring and defining the problem.

Interestingly, the thing that kills progress on easy problems is disagreement. Disagreement that results from debate about alternative solutions that can easily be measured and compared reflects an amazing tendency to take something easy and make it hard by arguing in order to justify our work. Conversely, the thing that kills progress on hard problems is consensus.

Although communication will be more important for hard problems, there will not be just one way to frame a hard problem, and solutions to hard problems—which are often complex problems—cannot always be evaluated in terms of solved and not solved. Therefore, debate about alternate solutions—which will not likely be routine or canned solutions—is superfluous. It is best to figure out what everyone can agree on, not how to get everyone to agree on everything.

Complex problems can intuitively be thought of in the following manner:

1) One person (often a skilled worker) can do it (neither quickly nor easily).

2) A group of people may be unable to work together to complete the task, to speed it up, or to add accuracy (sublinear performance improvements from horizontal scaling can at best be expected).

3) That group is unlikely to agree on the outcome or solution when complete.

A single individual often cannot perform well at a complex problem mainly because knowledge about the problem itself is not universally known, easy to acquire, or easy to share, thus making a complex problem hard to learn and harder to share. Consider jobs such as cybersecurity, intelligence analysis, conflict analysis, financial analysis, supply-chain risk, and medical diagnostics. These jobs cannot be completed quickly or easily, and they all require knowledge and years of on-the-job-training. Even then the "experts" may only know what makes their job complex but not how to remove the complexity, bias, or noise.[15]

Knowledge of complex problems is often minuscule because the knowledge is mostly fixed, whereas the problems are vast and constantly changing. This is why we tend to know what we know only after we need to know it. Polanyi's paradox asserts that we know more than we can tell, because what we know in complex domains includes tacit and implicit, rather than explicit, knowledge. The paradox is that for problems for which we need the most help, we are often in the weakest position to ask for it or to give it. If someone wants your help but cannot tell you how to help, then they are working on a complex problem. In fact, "the capacity of the human mind for formulating and solving complex problems is small compared with other problems."[16] Therefore, Polanyi's "true" paradox is that we often know less than we think we can tell and often think we know more than we can tell.

An indication that the acquisition of knowledge will be difficult to acquire and codify can often be found where the distribution of knowledge over workers is not uniform. For complex problems people cannot be easily swapped out for one another and will not be interchangeable, because universal knowledge of the problem does not exist. Thus many will work on only the parts they understand, and even within those areas performance may be inconsistent among workers. When the problem is not distributed, humans cannot be easily parallelized. Therefore, horizontal scaling with humans cannot be accomplished to solve a problem faster or with additional accuracy. Only vertical scaling works and only sometimes.[17]

---

15. For an example of noise, *see* Daniel Kahneman et al., "Noise: How to Overcome the High, Hidden Cost of Inconsistent Decision Making," *Harvard Business Review*, October, 2016, hbr.org/2016/10/noise.

16. H. A. Simon, *Administrative Behavior: A Study of Decision-Making Processes in Administrative Organizations*, 2nd ed. (New York: Macmillan, 1957), 198.

17. As a proxy for problem complexity just look at the distribution of compensation of workers currently solving a problem. If workers have an uneven distribution in pay, it may be because attempts to solve the problem are being attacked vertically with better-skilled workers. If pay is equal or roughly equal, then the problem is trying to be solved with horizontal scaling, where workers are interchangeable parts. When vertical scaling exists, complete automation will not.

The lack of uniform knowledge and uniform performance supports an earlier assertion that a human measuring stick is a bad idea for complex problems because any user represented as merely human would react inconsistently to some arbitrary solution. Extreme users often have problems that no average user has, and these users will tolerate just about any solution the day it is delivered. Because complex problems are often large problems, we will need to start somewhere. The extreme users with their extreme-use case is a great place to start. For simple problems there are no extreme users, because the average user is representative of all users.

Whereas simple problems often occur in closed domains with highly constrained types of data—which are narrow enough to explore extensively and be strictly evaluated—complex problems do not. Complex problems often occur in open domains; they adapt, change, and evolve. Complex problems are complex due to their scope (i.e., broadness), complexities (i.e., multiple tasks), ambiguities of an output (i.e., lack of universality and lack of certainty), and adaptive nature of inputs (i.e., features) over time.

Complex problems include:[18]

a) Multiple relationships among various entities. The complexity of the problem is usually represented by the sheer number of data sources, entities, and variables.[19] The more data sources, entities, and variables you have the more complex the problem. This is why solutions to complex problems are not always one solution but rather many solutions working together.

b) Hierarchical structure including the quantification of the order of tasks and the structure of how the problem is arranged in order of rank. If you are familiar with medical diagnostic hierarchies,

18. This mixture of features is similar to what is called VUCA (volatility, uncertainty, complexity, ambiguity) in modern approaches to management. *See* O. Mack et al., eds., *Managing in a VUCA World* (Switzerland: Springer International Publishing, 2016).

19. J. Funke, "Complex Problem Solving," in *Encyclopedia of the Sciences of Learning*, ed., N. M. Seel (New York: Springer US, 2012), 682–85.

cyberkill chains, or the MITRE ATT&CK framework, then you are already familiar with problems nested within larger problems.

c) Opacity involving the values of variables and labels. Hot dogs have little opacity. Although they share many characteristics with most sausages, which may make them difficult to describe, we routinely find hot dogs and skip Italian sausages at the grocery store. This fact results in variables and labels that are easy to evaluate. Defective potatoes are still potatoes, which introduces noise and thus some opacity, but we can still point out a defective potato rather easily. Medical imaging analysis like those that identify cancerous tumors are inherently opaque. After all, we don't examine X-ray images ourselves. This job requires experts to evaluate images. What constitutes a cyberthreat is inherently opaque. Opacity introduces noise and thus creates inconsistencies in solution performance. Opacity also leads to cause and effect being difficult to discern, understand, and predict. When cause and effect is difficult to discern, data does not speak for itself, rather we are speaking for it, and we often give it the wrong voice.

d) Complex problems highlight the role of time and changes in context of a problem that solutions must adapt to. Temporal dynamics create hidden relationships with intermittent temporal effects that exist between time-oriented events or interactions on various temporal scales on a variety of time lags.[20] Some time-oriented events move fast, and other time-oriented events move quite slowly. Slow dynamics in a system may dominate faster components of a system, though sometimes the fast changes the slow. Figuring out what is moving and how fast it is moving is part of problem comprehension because fast and slow are system constraints,

Cybersecurity, intelligence analysis, fraud detection, threat detection, finance, and medicine all have dynamic relationships,

---

20. That is, a period of time between one event or phenomenon and another.

often hidden, with intermittent temporal effects. For example, low-and-slow attacks, also known as slow-rate cyberattacks, involve what appears to be legitimate cybertraffic at a slow rate, making it difficult to distinguish from normal network traffic. This would be akin to a defective potato actively trying not to be a defective potato.

Researchers at the University of Toronto wanted to predict a patient's health over time, but data from medical records is messy.[21] Consider that throughout our lives we might visit the doctor at different times for different reasons at different locations, generating a smattering of measurements at different times and at arbitrary intervals. Conventional supervised machine learning struggles to handle intermittent temporal effects. The nature of complex problems requires solutions to learn across various stages of observation.

The challenge for problem solving in complex domains is having to accommodate numerous data sources, entities, variables, as well as hierarchical structures, opacity, and dynamic and intermittent temporal effects. Continuous learning (CL) is an area of machine learning that is useful when the very nature of a problem adapts from one learning episode to the next. By deploying a model with frozen parameters, continuous learning promotes a way to move a solution from the very rigid paradigm of training offline toward a smooth, flexible way of learning.[22]

Feedback loops are one way to accomplish CL, which occurs when a system's prediction feeds back into the training data in order to impact future performance. For example, a system may predict a cyberthreat, and

---

21. *See* Ricky T. Q. Chen et al., "Neural Ordinary Differential Equations," *arXiv*, June 2018, arXiv:1806.07366, http://arxiv.org/abs/1806.07366; it was crowned Best Paper at the Neural Information Processing Systems conference.

22. The traditional way in which models update their understanding of a problem is that we often retrain an entire model from scratch. In the meantime, our solution slowly falls off pace with the changes in the problem resulting in solution-rot. Retraining an entire model from scratch is called a cumulative strategy, which may be the best-performing method to prevent solution-rot, but it is expensive in terms of computational power.

a cyberanalyst may curate that output and provide feedback regardless of whether some prediction was right or wrong and thus tune the system for subsequent predictions. Or a system may recommend a movie that would subsequently influence future predictions. Feedback loops are what make our solutions wise.

Of course, solutions with feedback loops tend to be much more complex than solutions without—a fact that is often hidden because the phenomenon occurs when a solution is deployed.[23] In practice, feedback within a system is difficult to understand before deployment. That is, a solution with feedback loops tends to influence its own behavior and consequently its own performance.[24] Ultimately, feedback loops make solutions more vulnerable and less stable. Take the adversarial attacks on Microsoft's chat bot "Tay" for example. Adversarial data sets are input data sets designed to make models predict erroneously. They are data sets—either manually created or algorithmically generated—that stress test the consistency of a model and their outputs. Tay was taken down by Microsoft after the bot began generating sexist, racist, and offensive comments learned from manually generated user data.

All of this underscores an important point, which is that traditional software development often assumes that data is separate from the end solution. In machine learning, the data is the solution. However, data centricity adds other dimensions for errors that do not necessarily exist in traditional software development. The increase in complexity creates the

23. D. Sculley et al., "Hidden Technical Debt in Machine Learning Systems," *Neural Information Processing Systems: Proceedings of the 28th International Conference* 2 (Cambridge MA: MIT Press, 2015), 2503–11.

24. For example, the cause of the Flash Crash is disputed, but some analysts speculate that the Flash Crash was caused by erroneous signals, picked up by algorithmic trading programs, which then acted autonomously on these signals via nonexplicit feedback loops. A. Kirilenko, A. S. Kyle, M. Samandi, and T. Tuzun, "The Flash Crash: High-Frequency Trading in an Electronic Market," *Journal of Finance* 72 (2017): 967–98, https://doi.org/10.1111/jofi.12498.

additional requirements of debugging where solutions can fail on algorithm design and data engineering.

How Are Simple Problems Made Complex?

The concepts of simple and complex problems may be too fundamental in the sense that there is a vast amount of gray space between these two distinctions. It may be useful to understand some aspects of problems that influence difficulty. By manipulating problems, we may better understand the murky dichotomy and how problems move from simple to complex and back.

1) As a general rule of thumb, problem solving and good solution design becomes harder as the skill required for a job increases. Recall that Polanyi's paradox states that we often know more than we can tell. Thus, as a problem becomes harder, the need to be told what the problem is becomes more important, but it's harder to acquire because it will be harder to tell and thus harder to define. Problem solving is harder because problem definition and problem framing will be more difficult.

Additionally, as the skill required for a job increases so too does the cost of data acquisition and data labeling, because there are fewer people who can tell us what some value means. For example, we can crowdsource almost any group to identify a bad potato or a hot dog, even if it can't be explained why a potato is bad or a hot dog is a hot dog. But these same people will be unable to identify tumor in an MRI even after being instructed what to look for.

Furthermore, if you do not need to explain to someone why a potato is bad and can simply show them a bad potato as an output of a solution, then the solution does not require much or any interpretability or explainability. Interpretability is the degree to which a model's predictions can be explained natively. Explainability, meanwhile, is the extent to which the internal mechanics of a solution must be explained by another solution as the primary solution cannot be explained in human terms. It's easy to miss the subtle difference, but consider it like this: explainable solutions are opaque and require explanation post hoc, whereas interpretable solutions offer explanations native to the solution.[25]

There has been a recent explosion of work on so-called explainable AI where a second, post hoc solution is created to explain the first. DARPA has a program named Explainable AI, or XAI, which aims to produce more explainable machine learning solutions. Examples of "explainable" AI include saliency maps, which highlight where an artificial neural network was "looking" to inform a prediction, and it has been full of surprises that highlight how unintuitive machine learning can be.[26, 27, 28]

25. K. O'Rourke, "Explainable ML Versus Interpretable ML," https://statmodeling.stat .columbia.edu/2018/10/30/explainable-ml-versus-interpretable-ml/.

26. Saliency maps have grown in popularity because it is not always clear why an artificial neural network is making the predictions it is making. For example, you could use an artificial neural network to classify letters, but you will not find rules inside of the network that explains why an "a" was classified as an "a" and not a "B." Part of the complication is that artificial neural networks automatically create their own features for learning and also learns its own classification criteria. Blay Whitby explains in his book, *Artificial Intelligence: A Beginner's Guide* that the "great problem is that it may not be the one [feature] you want!" Whitby provides the example of an artificial neural network identifying the most reliable single feature which enabled it to classify tanks. The single feature however was the angle of the shadows in the foreground and background of the photo. Consequently, the solution predicted tank because it was the morning and predicted no tank because it was the afternoon. While this was an extremely effective way of classifying tanks it most certainly was not a program that recognizes tanks. Blay Whitby, *Artificial Intelligence: A Beginner's Guide* (London: Oneworld Publications, 2008), 67, Kindle edition.

27. Saliency maps still tell us little about what's actually going on with the internal workings of an artificial neural network. *See* Karen Simonyan and Andrew Zisserman, "Very Deep Convolutional Networks for Large-Scale Image Recognition," in 3rd International Conference on Learning Representations, ICLR 2015, San Diego, CA, May 7–9, 2015, Conference Track Proceedings.

28. Open-source projects include Skater, LIME, SHAP, Anchor, and TCAV. TCAV is noteworthy as the research attempts to deliver explanations in ways humans communicate with each other. While interpretability methods show importance in each input feature (e.g., pixel), TCAV instead shows importance of high-level concepts (e.g., color, gender, race) for a prediction class, which is more closely aligned with how humans communicate. *See* https://github.com/tensorflow/tcav.

Cynthia Rudin, professor of computer science at Duke University's Prediction Analysis Lab, describes these efforts as not going far enough.[29] Post hoc explanations are often unreliable and can be misleading. For complex problems it is often best to design solutions that are intrinsically interpretable, because they provide their own explanations, which are faithful to what the model actually computes. Deep learning, for example, is not as interpretable as shallow, simpler models, and shallow, simpler models are often less interpretable or explainable than rules, and they often rely on other solutions to explain outputs in human understandable terms.[30]

The point is, as a job requires more skill, additional demands are placed on communicating the output of a solution to others. When a solution has to communicate results, the problem is harder to solve because of additional demands to make solutions interpretable or explainable. If we think about it, this is Polanyi's paradox in reverse: the more a user knows about a problem (because they have to know more), the more they will want someone else to tell them about what they already know.

If a potential user has to know a lot about their problem, even if they can't tell you everything (or it will take years to tell you everything) about their problem, they will want your solution to be transparent in what it knows and how it knows it. It is not enough to provide just an answer. It may not be fair to a solution, but it makes sense on an intuitive level to most of us. After all, this is how trust is built between solutions and users, and it communicates to customers that you care enough to provide more than any old answer.

This tension between answers and explanations is another reason why GOFAI is often in the background of many real-world solutions. The reason is that customers do not care exclusively about answers. They care about stable, and often interpretable (or at least explainable), results. Therefore,

29. C. Rudin, "Stop Explaining Black Box Machine Learning Models for High Stakes Decisions and Use Interpretable Models Instead," *Nature Machine Intelligence* 1, (2019): 206–215, https://doi.org/10.1038/s42256-019-0048-x.

30. In fact, one of the best ways to understand a complicated model is to approximate a simple one. If you can do that successfully, use the simple one.

having a more stable system operating in the background or foreground provides a more stable value proposition, and often a more interpretable solution to customers.

2) The effect of knowing what a solution knows and how it knows it is magnified as the cost of an error goes up, because accountability goes up as well. That is, when the cost of an error is high, there will be someone accountable for that error. Therefore, concepts like safety, security, and interpretability all increase. Insiders will argue that human decision making is rarely explainable, difficult to interpret, and often functions as some post hoc argument. The implicit message, however, is that insiders do not want to be bothered with concerns that are at odds with their goals. Their goals typically include faster training and higher accuracy and certainly not friction created by interacting with flesh-and-blood people who they ultimately seek to replace. The truth is that humans understand the consequence of the loss of agency. So when cost of a mistake goes up, so too do the demands to be responsible. This effect increases the cost of design, which is further increased when some competing solution—like humans— already performs the job well.

My point: not all errors are equal, which means not all problems are equal. As failure becomes more expensive, the problem becomes more complex. Consider that false positives and false negatives for identifying bad potatoes is not the same in a cybercontext, which is not the same as in self-driving cars on public streets, which is not the same in large-scale kinetic warfare using autonomous agents. All problems have different costs, and all costs are compared to different competing solutions. As failure becomes more expensive, the problem becomes more important to solve completely. And when your solution has to be perfect, your problem is much harder to solve, because your solution will be much harder to design. The inverse is true. For problems that can be good enough, perfect is your enemy.

3) Closed problems will always be easier to solve than open problems. In fact, almost any solution in a closed domain will perform well, and almost any solution in an open domain will perform inconsistently.

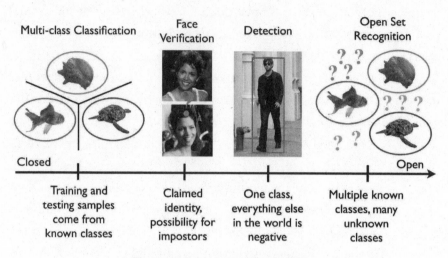

Figure 8.1 Open Set Recognition

Professor of computer science at the University of Notre Dame Walter Scheirer arranges (Figure 8.1) computer vision problems—though this is not restricted to computer vision problems—in order of their openness.[31] On the left are completely closed systems. Completely closed systems are rare in the real world. Benchmark data sets, for example, are synthetic problems and create closed domains where training and testing a solution occurs with a complete set of possible outcomes from known classes.

Recall Kewpie. Although their problem is real, it is not exactly found in the real world, and so it has closed characteristics. Hot dogs and not-hot-dogs are a one-versus-all problem. One-versus-all in binary classification is where we have one class that represents one object and another class that consists of all other objects. This problem is more open than predicting good and bad potatoes and thus more complex due to the ambiguity of "all."

---

31. W. J. Scheirer et al., "Toward Open Set Recognition," *IEEE Transactions on Pattern Analysis and Machine Intelligence* 35, no. 7 (July 2013): 1757–72, doi:10.1109/TPAMI.2012 .256.

Open domains include those domains in which we do not have knowledge of all possible outcomes. For example, self-driving solutions will often encounter scenarios that cannot be accounted for, and therefore modes of failure will exist that are completely psychologically implausible. Another domain with completely open problems is cybersecurity. Although we can train a solution to learn cyberthreats, there will always be an outcome that we cannot account for in training, so deployment will be complicated by unknown events. When deployed these solutions always respond differently to what is often an open domain, which is why we must always be skeptical of our training and test data and leery of whether it can ever represent the real world. This is why real-world solutions will come from a variety of places, including data, rules, knowledge bases, various analytical families, tribes, and learning paradigms. Again, this is also why GOFAI is in the background of many real-world solutions.

When thinking of open domains, it may be useful to think of the different "tails" of distribution of object occurrence for a problem. Short tails refer to common events. Fat tails include moderately extreme events that are more likely than you might expect. Long tails include extreme outcomes that are rare and often matter. Long tails are long because of edge cases, rare events, and obscure object occurrences. This complexity leads to many objects occurring infrequently and many new objects that have never been seen. Never-before-seen objects are the foundation of open set recognition problems for which there are not only many known classes but also many unknown classes.

The open set recognition problem is not well addressed by much of machine learning, because open set recognition requires strong generalization to unseen cases. The real issue is not that machine learning does not naturally generalize well—which it doesn't—but rather machine learning will still take a guess even when it is not confident in its prediction. If the observation is not actually represented as a class and does not have data to support that prediction, the solution will make a guess. Basically, supervised machine learning learns what are called class boundaries. These boundaries are a kind of bucket that the solution puts new, unseen data into. Most all supervised learning works in this way. Open set problems exist when the bucket is too large for the data and the solution indiscriminately throws the wrong items in the wrong, oversized buckets.

Ultimately, our solutions do not know they're performing in the real world with real-world distributions and real-world consequences. The first step for open set problem solving is to understand that you have one and not restrict training and testing to known classes. There are various strategies for open set recognition, but they generally seek to identify when a machine learning algorithm is guessing, which is effectively relegated to identifying low-confidence predictions.[32] In this case, a solution may reject low-confidence samples or reject unknown classes. Conversely, you can make your problem smaller, but there is always the risk that we make our problem so small as to make it fake. Another viable strategy is human-centric design in which the degree of collaboration between the human and the machine is predicated on the amount of uncertainty within the system itself. As uncertainty increases or cost of an error increases, the human centricity of design should increase as well.

4) Concept drift refers to the lack of stasis in a problem or the amount of change in the relationship between input and output data in the underlying problem over time.[33] For simple problems the mapping between inputs and outputs is static, meaning that the learned relationship is just as valid today as the relationship tomorrow. For example, a hot dog is always a hot dog. Conversely, complex problems evolve over time and solutions may become obsolete over that same period of time.[34]

---

32. Chuanxing Geng, Sheng-jun Huang, and Songcan Chen, "Recent Advances in Open Set Recognition: A Survey," *arXiv*, November 2018, arXiv:1811.08581, http://arxiv.org/abs /1811.08581.

33. Concept drift may also be called covariate shift, data set shift, or nonstationary, but the concept is different than data shift.

34. I. Žliobaitė, M. Pechenizkiy, and J. Gama, "An Overview of Concept Drift Applications," in N. Japkowicz and J. Stefanowski J., eds., *Big Data Analysis: New Algorithms*

Therefore, plasticity becomes important as a problem shifts. However, much of machine learning forgets as problems shift. For example, catastrophic forgetting is the tendency of a solution to completely and abruptly forget previously learned information upon learning new information. When a machine learns from data, the function changes slowly then abruptly, thus older information learned is lost.[35] It's widely thought that catastrophic forgetting is an inevitable feature of artificial neural network architectures, though some approaches remember old tasks by selectively slowing learning on the weights important to a problem.[36] Jason Brownlee, proprietor of Machine Learning Mastery!, explains various ways to handle drift, including (1) doing nothing, (2) periodically refitting your solution with new data, (3) periodically updating your solution, (4) weighting data, (5) learning the change, or (6) choosing a new solution.[37]

Data drift is related to concept drift. This is typically explained as a mismatch between the data used for training and testing and the data a solution encounters in the world. For example, a solution trained to spot signs of disease in high-quality medical images will struggle with blurry or cropped images captured by a cheap camera in a busy clinic. Workarounds are very similar to Brownlee's suggestions such as periodically retraining solutions, refreshing a solution with weights, using adaptive ensemble methods, or using streaming models as new data arrives. The point here is that as stability of a problem decreases, the complexity of the problem

for a New Society. Studies in Big Data 16 (New York: Springer, 2016), https://doi.org/10.1007/978-3-319-26989-4_4.

35. This is especially true for artificial neural networks due to the iterative nature of learning using back-propagation.

36. James Kirkpatrick et al., "Overcoming Catastrophic Forgetting in Neural Networks," *arXiv*, December 2016, arXiv:1612.00796, http://arxiv.org/abs/1612.00796.

37. Jason Brownlee, "A Gentle Introduction to Concept Drift in Machine Learning," *Machine Learning Mastery*, August 12, 2019, machinelearningmastery.com/gentle-introduction-concept-drift-machine-learning/.

increases, where the plasticity of a solution must correspondently increase as well, which complicates problem solving.[38]

5) Another problem with supervised machine learning in particular is that it is sensitive to context or the background distribution of data. Consider the preamble to the Goldilocks story, which describes Goldilocks as lost in the woods. There would be no bears if Goldilocks had this problem in a city. Solutions are context dependent because our problems are context dependent, which is why we must understand not only our problems but also the context of our problems.

Consider the problem of potatoes on a conveyor belt. The context of the potato never changes. The problem always includes a potato on a belt. Also, consider Deep Blue, where the board and time controls are the context. Time controls represent the constraints of the problem that a real human chess player would have to negotiate, but these constraints are within a game, and the time and board are fixed. For self-driving cars both the objects and the context of those objects change. Bridges, for example, are often considered black boxes for autonomous cars,[39] because bridges lack many of the environmental cues present on roads that can prevent sensors from keeping the vehicle on track. Complex problems are complex due to changes in object occurrence, but they're further complicated by the context surrounding the objects. While object occurrence and

---

38. Problem drift fails to underscore just how tricky deep learning is in the real world. A team of forty researchers at Google led by research scientist Alex D'Amour recently explored the concept of underspecification. Underspecification attempts to explain why solutions perform well during testing, and similarly when compared, but are unstable in the real world. The research shows that the same solution, trained on the same data with the same performance, will perform differently in the real world due to random variations in the initialization process, where those difference are often unknown until it is too late. *See* A. D'Amour et al., "Underspecification Presents Challenges for Credibility in Modern Machine Learning," *arXiv*, November 2020, arXiv:2011.03395, https://arxiv.org/abs/2011 .03395.

39. Michael Luciano, "Five Flaws Holding Driverless Cars Back," *Electrical Engineering News and Products*, September 29 2017, www.ecnmag.com/blog/2017/09/five-flaws -holding-driverless-cars-back.

environment changes don't fool humans, they do fool our solutions that learn from data.

Consider cybersecurity, and specifically malware detection, in executables. Although there are many types of malware and countless variants, it is easier to detect than something like so-called insider threat. An *insider threat* is a malicious threat to an organization that comes from people within the organization (hence *insider* threat), such as employees, former employees, contractors, or business associates. The context of malware detection in executables is often static, existing in a hash or the binary patterns in an executable. But because executables have not changed much in thirty years, we know where to look.

This is also why adversaries attempt to change the packaging of executables using packers, which are programs that compress an executable. By wrapping an executable in another package, the adversary changes the way the executable file looks, which makes detection more difficult. Crypters, polymorphic malware, downloaders, droppers, and staged loading are all countermeasures that change context, thus challenging solution performance.

The context of an *insider threat*—specifically—and the detection of aberrant behaviors—generally—is different because the baseline behavior is different from user to user, organization to organization, and also different over time. In other words, everything is changing all at once. When everything is changing the problem is more complex.

6) Everything may be related to everything else, but nearby things are more related than distant things.[40] Therefore, the further in the future you want to predict something, the harder the problem becomes. For example, let's say we want to predict the likelihood of cancer in a patient based on precursors. The amount of time those predictions live out into the future will influence solution performance. The further forward the prediction, the less confident you will be, and the

---

40. W. Tobler, "A Computer Movie Simulating Urban Growth in the Detroit Region," *Economic Geography* 46, supplement (1970): 234–40.

harder it is to get feedback on solution performance and ultimately feedback to reinforce performance.

Instead of making "hero" predictions with various dynamic components, it may be wise to decompose the problem into smaller components and investigate what is changing and at what speeds. This is true for forecasts with long time horizons as well multiple algorithms for which making smaller, simpler predictions with short time horizons may be more optimal. Decomposing an environment into simpler subsystems or processes limits the number of inputs, making them easier to control for confounding factors. For example, the BIM 360 Project IQ Team at Autodesk takes this small prediction approach to areas that contribute to construction project delays. Their solution predicts safety and scores vendors and subcontractors on quality and reliability, all of which can be measured while a project is ongoing.

7) Recall that meta-learning has several key application areas where complexity measures are applied to support problem comprehension for either combining solutions for a better solution or recommending the correct solution. As a result, complexity measures are exploratory. These measures can be used to understand the peculiarities of many data sets and consequently the peculiarities of many problems.

The complexity of a problem (specifically supervised machine learning problems) can be attributed to a combination of three main factors:[41]

- Overlap of features and classes.
- Separability of classes and the complexity of the function as well as smoothness of the function that must be fitted to the data.
- Geometry and topology that capture the distribution and overall structure of the data.

---

41. A. C. Lorena et al., "How Complex Is Your Classification Problem? A Survey on Measuring Classification Complexity," *ACM Computing Survey* 52, no. 5 (September 2019): 1–34.

For example, it doesn't matter what machine learning algorithm is used if ambiguity is present in classes that cannot be distinguished using the data available. Recall a class denotes a set of items, like hot dogs or hamburgers, that have certain common characteristics. The ambiguity of classes occurs when the problem is poorly defined or poorly conceived with the use of nondiscriminative features.[42] A discriminative feature is a feature whose values shows little overlap among the different classes thereby making the feature good at predicting a new object to an existing class, one in which features can discriminate among other classes.[43] For instance, a discriminative feature for predicting hot dogs doesn't also predict hamburgers.[44, 45]

The volume of the overlapping distributions of feature values within classes can also determine problem complexity by finding minimum and maximum values for each feature in the classes. The higher the value, the greater the amount of overlap between classes. The greater the overlap, the greater the problem's complexity due to the fact that the problem is interwoven, which also means the problem is harder to solve. If there is at least one nonoverlapping feature, then the value of the complexity measure should be zero and the problem easier to solve.[46]

---

42. In machine learning, a feature is an individual measurable property of the problem being solved.

43. Discriminant ratios the class overlap according to a single feature. This measure computes the ratio of inter-class to the intra-class scatter for each feature. *See* https://www.researchgate.net/figure/Example-of-F1-computation-for-a-two-class-dataset_fig1_335810225.

44. This measure computes the ratio of inter-class to the intra-class scatter for each feature (e.g., Fisher's discriminant ratio) Image: https://www.researchgate.net/figure/Example-of-F1-computation-for-a-two-class-dataset_fig1_335810225.

45. Low-discriminant ratio values indicate that there is at least one feature whose values show little overlap among the different classes. This measure indicates the existence of a feature that is distinct from other features. Therefore, it can separate classes fairly easily. Lower measures indicate a problem that is inherently easier to solve.

46. There are also various complexity measures that determine feature correlation. If at least one feature is highly correlated to the output, this indicates that simpler functions can be fitted to the data. *See* A. C. Lorena et al., "Data Complexity Meta-Features for Regression Problems," *Machine Learning* 107 (2018): 209–246, https://doi.org/10.1007/s10994-017-5681-1.

Complex problems are reflected in the complexity of the class boundary itself. Recall that machine learning is, more or less, fitting a function to data to find patterns or regularities. Specifically, supervised machine learning is using class boundaries to partition data into classes or buckets to predict class assignment. The class boundary is evaluated against the number of misclassified objects, where the goal is to minimize the misclassification rate. When there is some regularity in the data and little ambiguity between classes, a compact representation can be obtained. The complexity of a problem increases—that is, it is more difficult to solve—as more complex boundaries are required.

In other words, problem complexity can be understood by the shape of a decision boundary and characterization of the class separation. By understanding the decision boundary and class separation, a problem can be understood by how identifiable new instances can be classified in a supervised machine learning paradigm. For example, data points on the edge of a decision boundary surrounded by other data points are made more complex when data points are surrounded by other data points in different classes.[47] What that means is that a problem is harder to solve if our solutions have to be more complex to separate classes due to a lot of overlap between classes. Examples include distinguishing between Chihuahuas and muffins, Labradoodles and fried chicken, sheepdogs and mops, puppies and bagels, Shiba Inus and marshmallows and perhaps even hot dogs and sausages.[48]

Research has also shown that problems like fraud detection, network intrusion detection, and character recognition have different instances, many with very similar characteristics. When you have many classes that are very similar, there is overlap between classes in the so-called feature space. The class overlapping problem has become one of the toughest problems in machine learning and data mining. Researchers

47. See image at https://www.researchgate.net/figure/Example-of-overlapping-region _fig2_335810225.

48. Mariya Yao, "Chihuahua OR Muffin? Searching for the Best Computer Vision API," *Topbots*, September 22, 2017, https://www.topbots.com/chihuahua-muffin-searching-best -computer-vision-api/.

have found that misclassification often occurs near class boundaries, where overlapping usually occurs, and it is hard to find a feasible solution for the complexity.[49, 50, 51]

Complexity measures can be used as a rough guide for the grouping of solutions in what is sometimes called the domains of competence. Rules are extracted from the measures to identify when solutions can be expected to achieve good or bad performance on a given problem characterized by the data.[52] Essentially, complexity measures highlight that our

---

49. C. L. Liu, "Partial Discriminative Training for Classification of Overlapping Classes in Document Analysis," *International Journal on Document Analysis* 11, no. 2 (October 2008): 53–65; V. Garcia et al., "Combined Effects of Class Imbalance and Class Overlap on Instance-Based Classification" (conference paper, IDEAL 06, Proceedings of the 7th International Conference on Intelligent Data Engineering and Automated Learning, September 2006), 371–78.

50. Results show that the degree of class overlap has a strong correlation with class imbalance. An imbalanced data set means instances of one class are higher than instances in other classes. For example, a disease data set in which 0.01 of examples have positive labels and 0.99 have negative labels is a class-imbalanced problem, whereas a fifty-fifty split of positive and negative labels is *not* a class-imbalanced problem. Studies have been done on handling overlapping regions for the real-world data sets. See, for example, H. Xiong, J. Wu, and L. Liu, "Classification with Class Overlapping: A Systematic Study," The 2010 International Conference on E-Business Intelligence, 2010.

51. Problem-solving schemes including discarding overlap regions, merging overlap regions, and separating overlapping regions to model the data sets with the presence and absence of class overlap. H. Xiong et al. found that modeling the overlapping and nonoverlapping regions separately was the best scheme for solving the class overlapping problem. *See* H. Xiong, "Classification with Class Overlapping."

52. The knowledge advent from the problem complexity analysis can also be used for improving the design of existent solutions. For instance, researchers have proposed modifications to the back-propagation algorithm for training artificial neural networks that embed their concept of hardness. In this example, the error function of the backprop algorithm places more emphasis on hard instances of the problem. *See* Michael R. Smith, Tony Martinez, and Christophe Giraud-Carrier, "An Instance Level Analysis of Data Complexity," *Machine Learning* 95, no. 2 (2014): 225–56. Other works along this line include artificial neural networks researchers who use the measures class overlap and imbalance ratios (*see* Piyanoot Vorraboot et al., "A Modified Error Function for Imbalanced Data Set Classification Problem," in Proceedings of the 7th International Conference on Computing and Convergence Technology, ICCCT'12, 854–59) and boosted decision trees researchers who use neighborhood measures and feature efficiency measures (*see* Yoisel Campos, Carlos Morell, and Francesc J. Ferri, "A Local Complexity-based Combination Method for Decision Forests Trained with High-Dimensional Data," in Proceedings of the 12th International Conference on Intelligent Systems Design and Applications, ISDA'12, 194–99).

problems are not all created equally, it is not easy to know beforehand what the best solution is, and there is not one best solution for all problems. Rather, there are a variety of solutions, and your solution may be not be one solution but rather a complex array of partial solutions, each conditioned on some subproblem.

Is a Problem Impossible?

When humans are unable to solve a problem (or solve a problem consistently), our technological solutions will struggle, too. A human often has to be able to solve a problem before a machine can solve it, or solve it well before a machine can solve it well. Moreover, just because a human can do something, does not mean that a machine can. In fact, although machines can do things in ways humans can't, like repeating a task without a break and often with fewer errors, or do something faster, they can do almost nothing that a human can't at least do once, slowly, or poorly. A common myth is that AI will solve all problems that humans can and cannot solve, don't have, can and cannot see, and cannot describe. For this reason, an impossible problem occurs when:

1) One person cannot do it.
2) A group cannot do it better or faster.
3) No one can agree on the solution.

Nearly impossible problems are known as wicked problems because they persist in spite of the time and resources required to make any impact on them. Wicked problems have at least one, but often many, of the following characteristics:

No problem framing: Although complex problems will not have a well-defined problem, it will not be possible to create a well-defined problem statement for a wicked problem. The lack of problem framing is a problem for all wicked problems.

No stopping rule: Optimal stopping theory has been studied extensively in the fields of applied probability, statistics, and decision theory. The question is about the optimal strategy—or stopping rule—to maximize the probability of selecting the best solution. However, we cannot tell when we've found a solution for a wicked problem as the search for solutions never stops.

No trial and error: It's nearly impossible to determine if a solution is working until it is too late. That is, almost every solution in a wicked domain is a single-shot operation, and there are fewer opportunities to learn by trial and error. We get one attempt to solve wicked problems, which counts significantly, and the consequences cannot often be undone. Solutions to simple and complex problems can be easily tried and abandoned. There is no failing fast for wicked problems because there are no do-overs.

Every solution is a candidate solution: Ordinary problems (i.e., simple and complex) come with a limited set of potential solutions that can, to varying degrees, be evaluated as solved and not solved. However, wicked problems have countless solutions because the problem has no correct framing. Therefore, solutions are not solutions as much as they are trade-offs.

No distinction: Ordinary problems share characteristics with similar problems, so they can be approached in similar ways. Wicked problems are often unique and without precedent, so there's no experience to help the design of a solution. There are no qualifications for these problems, and no one is truly qualified to solve these problems.

The problem is really a symptom of another problem: In other words, most problems are self-contained, but wicked problems are entwined with other problems, and those problems don't always have a root cause.

Everyone is a customer: Because wicked problems can be explained in numerous ways, and impact many people, they will involve many stakeholders and shareholders. Needless to say, solving intelligence is a wicked problem because everyone is a customer. Self-driving technology is also wicked for various reasons, but mainly because everyone is a customer either directly or indirectly. When everyone is a customer, your solution has to be perfect, and when your solution has to be perfect, because everyone is a customer, there is a low tolerance for trial and error.[53] Having no room for error, and limited ability for trial and error, means it will take decades to perfect.

No room for error: When you can't be wrong, can be held liable for the consequences of your solution because solution performance will have such a large impact, and will have to justify to everyone your solution, then your problem is wicked, and you will spend decades on a solution. This is part of the reason why, when self-driving technology pundits tell you to pay attention in the next few years, you can look away and not worry much for a few decades.[54]

If you find yourself working on a problem with wicked characteristics, then you should consider changing your problem to fit what will invariably be an imperfect solution. For example, Chinese technology firms are turning the streets themselves into something that their self-driving solutions can handle.[55] There is no techno-optimism in China. It seems that when they don't like the problem, they change the problem. This may be a viable option for others, too. If solutions are good in game form, then figure out how a problem can be gamified. If solutions don't excel in open domains, why not, as China is doing, make your real-world problem less real by

53. For example, Courtney Linder, "Why Hundreds of Mathematicians Are Boycotting Predictive Policing," *Popular Mechanics*, July 24, 2020, www.popularmechanics.com /science/math/a32957375/mathematicians-boycott-predictive-policing/?utm_source =pocket.

54. This is exactly what Tim Adams suggested in his 2015 *Guardian* article: "Self-driving Cars: From 2020 You Will Become a Permanent Backseat Driver," September 13, 2015, https://www.theguardian.com/technology/2015/sep/13/self-driving-cars-bmw-google -2020-driving.

55. This approach standardizes the environment, so that inflexible machines may function semi-autonomously. "Chinese Firms Are Taking a Different Route to Driverless Cars," *The Economist*, October 12, 2019, www.economist.com/business/2019/10/12/chinese -firms-are-taking-a-different-route-to-driverless-cars.

making it a closed problem? China is focusing on the constraints of the problem rather than exclusively focusing on the limitations of solutions.

The point is that these characteristics are how problems are wicked or can become wicked if we are not careful. Virtually any problem can be impossible to solve if we expand the problem to an unsolvable dimension by unnecessarily including wicked characteristics. Make your problem as large as possible to acquire a customer. Perhaps increase the size for future customer acquisition. But do not make it so large as to make every solution inadequate. If your problem is intrinsically wicked, then good luck, but don't make a manageable problem wicked by making it *too* large to solve.

What Is a Problem Statement?

Herbert Simon, Amos Tversky, and others explain in a seminal 1987 paper that the three activities that constitute problem solving are: (1) choosing issues that require attention, (2) setting goals, [and] (3) finding or designing suitable solutions.[56] In other words, problem solving involves noticing problems, framing problems, determining how best to solve a problem, how much of it to solve, doing needed analyses for solution design, and ultimately developing a solution.

A problem statement is a description of a problem. A concise, well-written problem statement allows everyone involved to agree on the problem that is being solved. Ill-defined problems are many times more difficult to solve than well-defined problems. An ill-defined problem is better seen as a phenomenon for which it is possible to articulate or frame the problem in several ways.[57] This is because an ill-structured phenomenon includes incomplete information. When uncertainty and ambiguity

56. Herbert A Simon et al., "Decision Making and Problem Solving," *Interfaces* 17, no. 5 (1987): 11–31, www.jstor.org/stable/25061004.

57. C. Madsbjerg and M. B. Rasmussen, "An Anthropologist Walks into a Bar . . ." *Harvard Business Review* 92, no. 3 (2014): 80–88; V. Ramaswamy and F. Gouillart, "Building the Co-Creative Enterprise," *Harvard Business Review* 88, no. 10 (2010): 100–109.

make a problem difficult to grasp fully, people ultimately see different things in the same phenomenon.[58]

As mentioned earlier, Kolmogorov complexity suggests that the smallest solution needed to describe a problem is the least complex. Kolmogorov complexity also suggests that the smallest problem description needed to describe a problem is the least complex. In other words, the longer your problem statement, the harder the problem solving. If you can write the problem down and make it concise, you will in effect reduce the problem complexity. Ultimately, by understanding a problem, we make it manageable by making it as concise as possible. Intuitively, we can imagine a general rule of thumb to be: the easier a problem is to explain, the easier it is to solve; or the more work done in understanding and framing a problem, the easier it will be to solve. Well-defined problems—which are invariably shorter problem descriptions—are less complex than an ill-defined problem, because the length of the problem description will be less ambiguous.

How much time you spend will depend on the time-fidelity trade-off you're willing to make, which, of course, is influenced by the problem complexity. Look at the problem from different perspectives from different people. Seeing the problem with different eyes is a great way to have instant insight on new, often overlooked directions and potentially perilous traps. View the problem from the perspective of various customers, future customers, from the business, and the market. How does the competition see the problem? How do politicians see it? How about your mother? Your spouse? How about the person with a different background? Imagine how people in various roles would frame the problem.

Rewriting problem statements many times, each time using one of these different perspectives, is the first step to solution design. The goal in these early steps is to eliminate unnecessary information by reducing the size of the problem statement. If the problem statement is a page, reduce

---

58. P. M. Leonardi, "Innovation Blindness: Culture, Frames, and Cross-Boundary Problem Construction in the Development of New Technology Concepts," *Organization Science* 22, no. 2 (June 2010), https://doi.org/10.1287/orsc.1100.0529.

it to a paragraph. When it is the length of a paragraph reduce it to a couple of sentences and then to one sentence or as small as you can make it. Remember that almost all dysfunctions of problem solving can be found in communication, and the problem statement is an important mechanism to communicate the problem. By making the problem statement clear and concise, you will in effect reduce the complexity of the problem and reduce the risk of communication failures.

There is something magical about writing down a problem. It's almost as though by writing about what is wrong, we start to discover new ways of making it right. Writing things down will also remind oneself and our teams of the problem and the goal. Once a problem is written down, don't forget to come back to the problem statement. It is a guide. Problem solving often starts with great intentions and alignment, but when it counts most—when the work is actually being done—we often don't hold on to the problem we set out to solve, and that's the most important part of problem solving: what the problem is and why we are solving it to begin with.

Furthermore, do not needlessly seek out complexity by making larger solutions to solve needlessly bigger problems. Complexity bias is the logical fallacy where we find it easier to seek out complex solutions rather than a simple one. Without a problem statement, solutions tend to become more complex and expand to fill in the available time we've allocated for problem solving. Parkinson's law, named after Cyril Northcote Parkinson, states that "work expands so as to fill the time available for its completion." This is a sort of solution sprawl, similar to the urban sprawl that expands to fill in geographic spaces immaterial to how well the urban landscape serves it citizenry.

Lastly, problem statements are part of the solution, but they should not include any mention of a solution. After all, it is a problem statement, not a solution statement. At this point you should not care what your solution looks like or what its name is. We can't understand the problem if we conflate it with the solution. Solution-centric problem solving is not problem solving, and meaningless communication about solutions denies our eyes what we know to be true.

Questions to ask about your problem statement:

- What is the problem?
- Is it the right problem to solve given other problems?
- Is it the right problem for you to solve?
- When does the problem occur?
- Where and under what constraints does the problem take place?
- Why is it a problem?
- How does the problem impact the customer?
- Is the customer better with the solution than without?
- What amount of change will be required?

Decompose the Problem

If you struggle creating a problem statement, then be prepared to decompose the problem. Reductionism involves breaking down a problem into smaller parts—and into smaller problems—that are more manageable and easier to understand and solve. Reductionism is necessary when we can't solve problems directly. Often, smaller is better. In fact, in nearly all domains, we should start with something small that can stand on its own two legs and expand from there.

Although big problems tend to allure top managers to spend time, they don't inherently require big solutions to better manage them. In fact, a persistent problem with big problems is that they attract big thoughts, big solutions, and big budgets that cost even more than the hourly rate. As Michael Schrage says in *The Innovator's Hypothesis*, "big problems become beacons for big budget consultants, thinkers, and advisors who want to launch comprehensive studies and holistic initiatives."[59] Bigger is seldom best and rarely cost-effectively better.

For complex problems specifically, reductionism is almost certainly required. Gall's law states that "a complex system that works is invariably

---

59. Michael Schrage, *The Innovator's Hypothesis* (Cambridge, MA: The MIT Press, 2014).

found to have evolved from a simple system that also worked." The corollary is that a complex solution designed from scratch rarely works and cannot often be made to work. Because large, complex problems are often too complex to understand in whole, the right solution is often grown with smaller batches rather than built as one solution.[60] Simple solutions to complex problems may be wrong, but complex problems cannot be solved with initially complex solutions. Gall's law suggests that we have to begin with the smaller, simpler solutions and smaller and simpler problem-solving processes that work and then scale with additional components and additional steps.

René Descartes introduced reductionism as the study of the world as an assemblage of physical parts that can be broken apart and analyzed separately. Descartes proclaimed: "Divide each difficulty into as many parts as is feasible and necessary to resolve it." Specifically, he is referring to his method for evaluating the logic of a statement, but the applications are much broader, and more useful for problem solving. So often we're weighed down by the magnitude of a problem, uncertain which direction to take or if we can get ourselves unstuck at all. The "difficulty" of a problem is "rarely a single challenge but rather a culmination of smaller tensions. The way to address these tensions is one at a time. If at any point you get stuck, break the problem down further."[61]

If there were a user's manual for reductionism, it might be presented as follows: Explore the whole of the problem, even if this view is still not a clear one. Pose an interesting question about part of the problem. Think out alternative, conceivable answers. If too many conceptual difficulties are encountered, back off and search for another aspect of the problem. When you finally hit a soft spot, search for experiments or further observation of the customer that can be easily conducted to provide answers so that no

60. Fred P. Brooks, "No Silver Bullet—Essence and Accident in Software Engineering," *Proceedings of the IFIP Tenth World Computing Conference* (1986): 1069–76.

61. Shaunta Grimes, "This 500-Year-Old Piece of Advice Can Help You Solve Your Modern Problems," *Medium*, Forge, December 11, 2019, forge.medium.com/the-500-year-old-piece-of-advice-that-will-change-your-life-1e580f115731.

matter what the result, the answer to the question will be convincing. Stay with the problem longer than anyone else. Become familiar, even obsessed, with the problem in its whole and in its parts. Love the details and the feel of all of the problem for its own sake. Use the result to find new questions and related problems. Every problem has some version that is large enough to be seen and small enough to solve.

Of course, you might go too large in problem specification, which will produce noise and introduce marginal information that draws attention away from the most relevant aspects of the problem.[62] That's okay. American physicist Murray Gell-Mann received the 1969 Nobel Prize in Physics for his work on the theory of elementary particles. Gell-Mann embraced an idea he named "plectics." The idea of plectics is going bigger and going smaller on problem specification. The lesson is that some problems are too small and precise to be interesting or too large to ever be solved. Problem solving is predicated on perception and ultimately on finding the correct focal point. That may require going up and down in the size and scale of the problem.

Each problem is a smaller piece of a larger problem or a large problem that is composed of smaller parts. Within stratifications of problems there are also adjacent problems that we may need to explore by moving around a problem laterally and vertically. If you feel you're overwhelmed with details or looking at a problem too narrowly, look from a more general perspective. To make your problem more general, ask questions such as: What's this a part of? What's this an example of? Or, what's the intention behind this? If the problem is part of a larger problem, it also means that each problem is composed of smaller problems, each more specific than the original, which

---

62. Rosabeth Moss Kanter argues that the best leaders can zoom in to examine problems and then zoom out to look for patterns and causes. They don't divide the world into extremes—idiosyncratic or structural, situational or strategic, emotional or contextual. The point is not to choose one over the other but to learn to move across a continuum of perspectives. R. M. Kanter, "Zoom In, Zoom Out," *Harvard Business Review*, 89 no. 3 (2011): 112–16.

may lend insights unavailable for the larger problem. Problem solving is all about finding the right level of abstraction.

Reductionism is not a cure-all, but it has lots of practical value for real-world problem solving as long as you keep returning to the real-world. Insiders avoid reductionism because the perceived epistemic value of learning about the self and intelligence cannot be reduced into smaller parts. Therefore, any attempt to solve it requires complete problem specification, holistic problem solving, and ultimately complete, end-to-end solutions.[63]

Furthermore, simple problems are not often compatible with reductionism because you may already be working in the smallest size of the problem. Although reductionism is a pragmatic methodology for approaching complex problems, it is not a solution for complex problems by itself. In other words, reductionism is a good place to start, but it's not the best place to finish.

Blind reductionism has a price, even for those problems that it is best suited to solve. Consider that going too small can cause important aspects of the problem to fall into blind spots. For example, complex solutions for which each smaller batch is solved and implemented separately will adhere to the approach known as "separation of concerns." In computer science, this is a design principle for separating a computer program into distinct sections such that each section addresses a separate concern. However, the separation of concerns effectively becomes a diffusion of responsibility when no one is concerned with the overall solution or is able to explain the whole of the problem or the entire solution. If reductionism creates the "separation of concerns," then that means no one is concerned about the overall problem or solution. We cannot get lost in part of the problem or part of the solution.

---

63. For example, embodied theory of cognition is a kind of holistic problem solving. Rodney Brooks, former director of the AI lab at the Massachusetts Institute of Technology, is one of the leaders of the "holistic approach" to artificial intelligence. Rodney A. Brooks, "Intelligence Without Representation," *Artificial Intelligence* 47, no. 1–3 (1991): 139–159, https://doi.org/10.1016/0004-3702(91)90053-M.

In practice, we will oscillate between parts of a problem and the whole of the problem and between reductionism and holism. The reason is simple: effective problem solving requires you to understand where to start and where to stop. We must figure out what a problem is, what it means, where it starts, and where it ends. These are important boundaries that all need to be understood because boundaries tell you what to do and what not to do. We will oscillate between the whole and the parts because if we can no longer see the consequences of our decisions at an atomic level, we may blind ourselves to the true size of the problem.

For example, point products provide a partial solution to a problem, rather than addressing all the requirements that might otherwise be met with a multipurpose solution. Problem solvers may constrain a problem to achieve performance gains but at the expense of adjacent parts of the whole of the problem. The result is what MIT professor Daron Acemoglu calls "so-so technology." So-so technology is an incomplete response to partially solved problems, which creates new problems, because partial solutions often bleed at the edges of adjacent problems. A partial solution without the support of other solutions will result in a solution for which the size of the underlying problem tears the partial solution apart.

Acemoglu explains that so-so technology doesn't often improve productivity because it ignores the whole problem. Consider that solving multiple, smaller problems is not equal to solving a large, complex problem with smaller solutions. The ability to switch between reductionist and holistic modes of problem solving will positively impact the performance and efficiency of your customers, not merely the performance of a narrowly defined solution. Do not be supplanted by a challenger who is working on parts of the problem that you decided to ignore or compete with incumbents on parts of the problem you didn't think were important to solve.

There are many point products in cybersecurity. These point products often align with specific threat vectors. By reducing the size of the problem, security vendors can build a solution faster, achieve performance gains easier, and go to market sooner. These point products have decomposed a problem to something small enough to solve and large enough to acquire a customer. Good for them. However, they have also left the vast majority of the underlying problem untreated. The fragmentation of both the cybersecurity problem and market is one of the reasons why Cybraics has focused on the whole of the problem while still using reductionism to understand each problem one at a time, but ultimately all at once from a customer perspective.

In 1945, mathematician George Pólya created a small volume describing methods of problem solving titled *How to Solve It*.[64] Pólya's suggestions include considering a problem from various scales or other angles, going bigger and going smaller, and asking how one problem relates to other problems. This is all in an effort to understand a problem from a variety of heights, various angles, and from various points of view. *How to Solve It* has sold more than a million copies and has been continuously in print since its first publication in 1945.[65] In fact, Marvin Minsky said in his paper "Steps Toward Artificial Intelligence" that everyone should know the work of George Pólya on how to solve problems (Table 8.1).[66]

---

64. George Pólya, *How to Solve It* (Princeton, NJ: Princeton University Press, 1945).

65. R. G. Dromey was inspired by Pólya's work. Dromey's book, *How to Solve It by Computer*, is a computer science book on problems. It contains a dictionary-style set of heuristics, many of which have to do with generating a more accessible problem.

66. M. Minsky, "Steps toward Artificial Intelligence," in *Proceedings of the IRE* 49, no. 1 (January 1961): 8–30, doi:10.1109/JRPROC.1961.287775.

Table 8.1: *How to Solve It,* by George Pólya

Heuristic	Informal Description	Formal Analogue
Analogy	Can you find a problem analogous to your problem and solve that?	Map
Generalization	Can you find a problem more general than your problem?	Generalization
Induction	Can you solve your problem by deriving a generalization from some examples?	Induction
Variation of the problem	Can you vary or change your problem to create a new problem (or set of problems) whose solution(s) will help you solve your original problem?	Search
Auxiliary problem	Can you find a subproblem or side problem whose solution will help you solve your problem?	Subgoal
Here is a problem related to yours and solved before	Can you find a problem related to yours that has already been solved and use that to solve your problem?	Pattern recognition Pattern matching
Specialization	Can you find a problem that is more specialized?	Specialization

Heuristic	Informal Description	Formal Analogue
Decomposing and recombining	Can you decompose the problem and "recombine its elements in some new manner"?	Divide and conquer
Working backward	Can you start with the goal and work backward to something you already know?	Backward chaining
Draw a figure	Can you draw a picture of the problem?	Diagrammatic Reasoning
Auxiliary elements	Can you add some new element to your problem to get closer to a solution?	Extension

If you can't define a problem, be prepared to decompose a problem. If you can't decompose a problem, be prepared to live with the problem until you can acquire the problem, or some part of the problem. By living the problem as the problem is lived, we can take something poorly understood and gradually add structure. Herbert Simon devoted part of his career to the subject of problem solving and says that what chiefly characterizes good problem solving from more mundane forms are (1) willingness to accept vaguely defined problem statements and gradually structure them, (2) remaining preoccupied with problems over a considerable period of time, and (3) extensive background knowledge in relevant and potentially relevant areas.

Living with a problem is akin to the "observation" phase that typically occurs within design thinking whereby designers watch how people behave and interact with their problem. Living with a problem is like the admonition to *walk a mile in someone else's shoes* before judging them. You must understand other's experiences. For problem solving, this could be framed as: before you solve someone's problem, live their problem. Living the problem in a similar environment with the same constraints is even better if you can. Always live the problem as the customer lives the problem. If you can't live the problem, then it's possible that there's no problem, or that the distance between you and the problem is too great for acquisition of it.

Go Small

Figuring out where to start is often the hardest part of problem solving. We get weighed down by the size of problems and which direction to take. When stuck, we wonder if we can get ourselves unstuck at all. As a problem becomes bigger, harder, and more complex, this is especially true.

Recall that Descartes reminds us that any "difficulty" is rarely a single challenge but rather a culmination of smaller tensions. Problem solving by reductionism shows a capacity to reason from simple, smaller, and often more useful principles to larger, more complex principles. In this way, going small with small batches of work are hedges for biting off more than we can chew. When you actually start the process of solving you should not abandon this practice. Everything should be made as simple as possible with smaller and simpler solutions.

If you don't know where to start, consider picking the smallest and easiest part of the problem with the smallest and easiest solution. This approach has the lowest cost, which will likely lead to determining other parts of the problem and eventually the most important part of the problem. There is a misconception that a solution can be too simple. It can't. Again, you're problem solving. Your solutions do not need to satisfy any external goal to be real. A solution cannot be too simple. The only way a solution can be shown to be too simple is if it is shown to be wrong. Similarly, a problem can only be too small after it has failed at acquiring a customer. Solving a problem is the goal, and simplicity is okay. In fact, as opposed to AGI, simplicity, reductionism, and incrementalism are all desirable characteristics for most all problem solving.

From a design perspective, working smaller and working toward simplicity is an effort to understand every part of a problem. The goal of going small and simple is the goal of designing and developing a Minimum Complex Solution (MCS). MCS is a search for the lower bound of complexity instead of the higher bound.[67] After all, the race for complexity is a race that cannot be won. The idea is to find the minimum solution and the minimum number of changes needed to create a solution where the problem does not exist. The goal of building solutions is not to find the most complex version of the problem or the most complex solution but the least complex versions.

Small batches work like Occam's razor. Occam's razor explains that given two solutions that perform more or less equally, you should always prefer the one that is less complex. For this reason, there will always be cases in which smaller, simpler solutions will be preferred to more complex solutions. Even if more complexity can manage to squeeze more accuracy, an inherently complex solution may require more time to build, train, debug, maintain, secure, and explain.

Consider the story of Netflix's one-million-dollar prize for the *best* possible recommendation engine, which was developed, but never implemented because it was too complicated.[68] Although saying yes to what is seen as the best is easy, it is also dangerous. Be careful about saying "yes"

67. S. Greiff and J. Funke, "Measuring Complex Problem Solving: The MicroDYN Approach," in *The Transition to Computer-Based Assessment. New Approaches to Skills Assessment and Implications for Large-Scale Testing*, F. Scheuermann and J. Björnsson, eds. (Luxembourg: Office for Official Publications of the European Communities, 2009), 157–63; J. Funke and S. Greiff, "Dynamic Problem Solving: Multiple-Item Testing Based on Minimally Complex Systems," in *Competence Assessment in Education: Research, Models and Instruments*, D. Leutner et al., eds. (Heidelberg: Springer, 2017), 427–43.

68. Ryan Holiday, "What the Failed $1M Netflix Prize Says About Business Advice," *Forbes*, April 24, 2012, www.forbes.com/sites/ryanholiday/2012/04/16/what-the-failed-1m-netflix-prize-tells-us-about-business-advice/.

to more than what is required. Ultimately, saying yes to the best needs a lot of justification if it is not obvious how much "yes" costs. I suggest saying maybe, but almost never yes, until you've done the homework to understand what yes costs and what trade-offs exist.

Keep in mind that the "best" does not exist in an absolute sense. The best is relative to your problem, given your constraints and resources. In unconstrained environments, there's always a better solution, but the real world always contains constraints. The best often means more. More expense, more complexity, more opacity, more development time and costs, and more deployment complexity and cost, all for a few extra percentage points, which is rarely noticeable to users.

Insiders, for example, have a strong desire to build a single, complete solution that is meant to handle all situations. However, ML-based solutions can barely be shipped as one complete algorithm. One of the reasons is that problems have more differences than we usually expect. Although end-to-end solutions handle difficult tasks like feature engineering and are meant to handle all situations, they are not required for problem solving. Real-world solutions are sloppy and are composed of many different sizes of working solutions. Moreover, when large-batch solutions have to be replaced, the whole solution has to be replaced, whereas smaller batches allow you to replace smaller pieces. Large-batch designers may not much care if the problem tears the solution apart, but small-batch designers do.

Moreover, simple solutions often require lower-skilled data scientists with less experience, whereas larger, newer, and more complex solutions often require higher-end data scientists.[69] Newer, complex solutions take longer because the skills to develop them may be harder to acquire. If you care about capital expenditures this point matters. Newer solutions are harder to make work in a production environment because they have not always seen reproducibility. And because few people have worked with the newest solutions, we are often in a very weak position to know the

---

69. This seems to be especially true for deep learning, which seems to work based more on anecdotal evidence, rather than hard and fast rules.

various modes of failure—most of which will be silent—and less prepared to respond when they do fail.[70]

Even when a simple solution is not your production solution, it often offers a powerful instrument to discover baseline performance and detect problems with your infrastructure (e.g., is the data where it needs to be to get the output of the solution where it needs to be) and data (e.g., getting data to the algorithm), all before production. Ultimately, there are many things to get right before you really want to think about the rightest solution. For example, "Google's Rules of Machine Learning" notes how some internal teams aim for a so-called neutral first launch.[71] A neutral first launch explicitly deprioritizes machine learning gains to avoid getting distracted. The motivation behind this approach is that the first deployment should involve a simple model with focus spent on building the proper data pipelines required for solutions at scale. This allows you to deliver value quickly and avoid the trap of spending too much of your time trying to "squeeze the juice."[72]

Consider picking some simple model that you could not possibly screw up, like a linear classifier or a very shallow convolutional neural network (or whatever is relevant given your team and problem).[73] Train the simple model, visualize losses and other metrics like accuracy, and model

---

70. The most frustrating part of software development is that it works, or it does not work. Humans are not like software programs. That is, we fail, but do so gracefully. Consequently, AI researchers seek solutions that fail gracefully. The concept of graceful degradation in software development described fault-tolerant systems. To be sure, machine learning fails much more gracefully than traditional software development. This feature is also a bug. Sometimes you want your solution to not fail gracefully. For example, this appears to be a solution that is failing to fail ungracefully: https://www.theverge.com/2021/5/14/22436584/waymo-driverless-stuck-traffic-roadside-assistance-video.

71. Martin Zinkevich, "Rules of Machine Learning: | ML Universal Guides | Google Developers," Google, June 12, 2019, developers.google.com/machine-learning/guides/rules-of-ml/.

72. Geoffrey Hinton, Oriol Vinyals, and Jeff Dean, "Distilling the Knowledge in a Neural Network," *arXiv*, March 2015, arXiv:1   arxiv.org/abs/1503.02531.

73. Andrej Karpathy, "A Recipe for Training Neural Networks," *Andrej Karpathy* (blog, April 25, 2019, karpathy.github.io/2019/04/25/recipe/.

predictions, and perform a series of ablation experiments with explicit hypotheses along the way. Ask questions and do it again.

Ablation experiments are when you vary one thing at a time on your solution.[74] In other words, when you work on a problem—any problem, but especially a complex problem—you should consider working not just in small sizes but also in small steps that facilitate learning aspects of your problem incrementally. Thus you'll save yourself the headache of deploying a large solution as a first step. For example, if you have multiple classes to plug into your classifier—say you want to add zucchini and other elongated-shaped foods to your hot dog food classifier—add them one at a time.

Always start small and take small steps to ensure that performance is what you want. Don't try to boil the ocean with the whole of a problem. With smaller steps almost everything can be reduced to something more manageable. Working in smaller sizes and smaller steps goes for your team as well. Rather than having your whole team work on something for six months, think about what one person can do in six weeks. The Basecamp team uses six weeks, which I think is a good size. If you are an Agile team, you may have batches of two weeks.[75] That is fine, too. The point is that constraining batch size will force everyone to find the best bad solution, rather than working into the abyss of perfection.

Of course, simple problems are different. Simple problems can often be solved by applying a single solution to the whole of the problem. In practice you may not know the best solution a priori. One strategy to find the best solution for a simple problem may be to simply guess. Guessing, however, will have a high error rate in the face of increasing complexity.

Alternatively, we could throw the kitchen sink at a simple problem, trying all solutions, or at least all solutions one can conceive. The kitchen sink model of problem solving is trial and error—a type of solution-guessing

---

74. Ablation helps. The more complex your solution, the less it will help. Deep learning especially is hard to debug by ablation due to the huge number of parameters interacting in nonlinear ways.

75. *See* https://agilemanifesto.org.

in which we conceive of every possible solution. The thinking goes that if you had twenty days to solve a problem, the kitchen sink model spends nineteen days trying different solutions and one day deciding what solution to select.

On the other hand, we could use meta-learning to streamline the trial-and-error approach to support the recommendation and selection of a solution and its configurations. After all, trial and error is very laborious and subjective given the many choices that need to be made, and MtL may prove to be more productive and efficient for selecting canned solutions. Trial and error is not as robust or defensible because it often lacks a conceptual understanding of a problem and its relationship with a solution. The issue with only knowing which solution is better is that you don't know why one solution is better. If we don't know why one solution is better than another in the face of increasing complexity, we will not know how to modify or create a solution to fit a problem.

Moreover, when a problem is so complex as to not be solved with a single, preexisting solution, then the trial-and-error approach fails to provide much value. Trial and error is about finding one solution, so a problem has to be small enough to be solved with one solution. Trial and error results in this solution *or* that solution. For complex problems, there are often cumulative problem-solving strategies. Cumulative problem solving is good at connecting the dots. Cumulative means connection by successive addition. Cumulative strategies are about finding all solutions. They look at this solution *and* (or *plus*) that solution. This means the use of partial solutions that learn how to best combine the results from other partial solutions.

For some technical context, examples of small, simple solutions include:

Rule-Based Algorithms: A rule-based algorithm is simply a formula given to a computer for it to complete a task as a series of rules. This is different from a machine learning algorithm, which "learns" or infers these rules, rather than having them explicitly stated as rules. If you know something about your problem, try to say it in a rule-based algorithm.

Linear Regression: Developed in the field of statistics, this is a model for understanding the nondeterministic relationship between input and output variables, specifically the linear relationship between variables. Simply put, linear regression draws a straight line (i.e., function) through data by minimizing error. Businesses use linear regression quite often—in fact, more often than any other model—to analyze the impact of price changes, assess risk, and evaluate sales estimates.

Logistic Regression: This is named for the function at the core of the method. Rather than the linear function in linear regression, the logistic function is an S-shaped curve—also called the sigmoid function—that can take on any real number and map it into a value between zero and one. A key difference between logistic (also called log-odds and logit) and linear regression is that the values being modeled are binary rather than numeric. Logistic regression is in fact a linear method, but predictions are transformed using a logistic function, which simply means that logistic regression is a sort of classification algorithm that predicts probabilities of classes.

Linear and logistic regression models are not what we think about when dreaming of AI. However, according to the 2019 Kaggle Machine Learning and Data Science Survey, the most common methods in use today are linear and logistic regression, with 80 percent of respondents claiming to use at least one.[76] Although not as powerful as more complex solutions, they can still produce useful results, are easy to use, cheap in terms of computation and implementation, and easy to interpret. Many high-performing solutions are built with these two methods as part of the solution stack, but they may be rejected by others who seek to make the simple, complex.

Decision Trees: This is a nonparametric method used for both classification and regression tasks. "Nonparametric" is fancy statistical talk for a solution that does not involve any assumptions. Decision trees are

---

76. "State of Data Science and Machine Learning 2019," *Kaggle*, 2019, www.kaggle.com /kaggle-survey-2019.

generally in the form of if-then-else statements, and tree structures can represent deep, complex rules.

Random Forest: This is a set of decision trees, and it considers the output of each tree before providing an *n* aggregate response. Each decision tree is a conditional classifier in which the tree is visited from the top, and at each node a given condition is checked against one or more features of the analyses data. These methods are efficient for large data sets and excel at multiclass problems, but deeper trees might lead to overfitting.[77]

Boosting: This technique iteratively combines several weak learners (small, shallow, and the not very accurate) into a strong learner (larger, deeper, and more accurate). The prediction after boosting is a weighted average of the predictions by multiple weak models.

Simple (or Shallow) Convolutional Neural Network (CNN or Convet):[78] CNNs recognize spatial patterns. It is the so-called convolutional layer that makes CNNs specialized for tasks like image processing. A simple CNN has one or just a few hidden layers, as opposed to something more complex like a multiscale Atrous Spatial Pyramid, Pooling Feature Pyramid, Network Residual Network (ASSP, FPN, ResNet). If you don't know what ASSP, FPN, and ResNet are, don't worry about it. In fact, it is probably best that you don't. The point is, avoid the most complex solution, if you can, in lieu of a simple one.

Making a solution complex rarely creates value. Over time a solution may become complex, but we should not seek complexity. Your goal should be to work off complexity, not to add it. Complexity bias is a logical fallacy that leads some of us to gravitate to complex solutions, but customers rarely want more complexity. More often, they want less. Many believe that we have to add AI or ML to a solution to justify our solution, justify our work, or substantiate our personal value. This is often misplaced, because

77. L. Breiman, "Random Forests," *Machine Learning* 45 (2001): 5–32. https://doi.org/10.1023/A:1010933404324.

78. Yann LeCun et al., "Gradient-based Learning Applied to Document Recognition," *Proceedings of the IEEE* 86, no. 11 (1998): 2278–324.

problems do not care how they are solved. In fact, my personal view is that this tendency is often overcompensation for the lack of understanding of a problem or an arrogance that a solution is more important than a problem or a customer.

Solve the Right-Sized Problem with the Correct Amount

Finding the right problem to solve is hard. Solving the right problem is harder. Solving the right problem to the correct amount is hard to know, so many assume when "doing AI" that you have to beat human measuring sticks.

A useful stopping rule for most of us occurs when a solution creates a customer. This is product-market fit. It does not mean you are done solving, but it means you have finally created a solution, because it has been positively evaluated by a customer. In other words, part of problem solving includes the goal of problem solving. The problem definition will outline what to change, why to change it, and what amount of change will be required. If you are unable to conceptualize the goal, then you will not know when to stop solving or what the trade-offs are for continuing.

How do you know how much to solve to create a customer?

You may not.

Of course, part of customer development is customer research, which is important for understanding whether you're solving a problem well enough to make a difference in someone's life. The more you can quantify how much better your solution needs to be to make a difference, the better: where "perfect" performance is not an acceptable answer.

In fact, try to find the minimum performance goals to validate with users and customers. Just like software start-ups launch when they have built a minimum viable product (MVP) in order to collect actionable feedback from initial customers, AI start-ups should launch when they reach the minimum performance required by early adopters. As noted earlier, this may not always be a machine learning solution. But if it is, the minimum performance is still the same. Remember, it doesn't matter how we deliver value, just that value is delivered.

Furthermore, performance requirements are best understood when they are customer-centric rather than solution-centric. In other words, being done is not defined by a release date. Being done is defined by the customer requirements and minimum goals. You can call your solution an MVP, but it does not make it so. Steve Blank highlights how traditional product development hides the market and hides the customer. Blank advocates for a customer development model that focuses on the customer, not the ship date or your first version of the solution.[79]

When the minimum performance requirements are easy to achieve, you will be able to launch quickly, but you may be continuously looking over your shoulder for copycats.[80] Harder-to-achieve minimum performance problems are harder to define and more difficult to accomplish, but they may also be harder to replicate. For example, solutions that can easily achieve high accuracy with inexpensive human labor may demand higher-performing solutions before they can find an early adopter. Although, it's just as likely that companies—as seen in the Kewpie example—may solve these problems themselves. Tasks requiring fine motor skills, for example, have yet to be taken over by robots because human performance sets a very high threshold to overcome. Projects attacking high minimum acceptable performance problems must invest more time and capital to launch.

Hero Problem Solving

Avoid the trap of being the hero problem solver by finding help. There is an illusion that problems are solved by one person. It is rare that we have all (or even most) knowledge of a problem, so finding domain expertise

---

79. Steve Blank, "The Customer Development Manifesto: Reasons for the Revolution," *Steve Blank* (blog), September 17, 2009, steveblank.com/2009/08/31/the-customer-development-manifesto-reasons-for-the-revolution-part-1/.

80. Ivy Nguyen, "AI ROI Tookits for Building AI Businesses," *AI ROI* (blog), roi-ai.com/.

from another part of your team or organization is crucial. Besides, complex problems will impact various people who have their own agendas and preferences. Therefore, including and excluding the right and wrong people will have positive and negative impacts on success.

The "right" people are obviously those who are familiar with the problem. That expertise, however, may not come from the office next door. More likely (especially in large organizations) the right expertise will come from some far corner of the business. Do not assume that the expert who is easiest to find is the best. Kevin Kelly has the Rule of Seven. The Rule of Seven effectively states that anything can found if you are willing to go to seven people. If you are willing to chat with at least seven people you will often get the right information.

As you talk to others create a matrix of the experts, the stakeholders, interested parties, noninterested parties (and why they're not interested), real users, data owners and stewards, and potential customers. The matrix will help you decide who plays what role and how, as well as what resources are available versus which need to be acquired.

The "right" people may include those outside of the organization. The right time to know whether the right person exists externally, however, is after you have lived with a problem, tried to live with a problem, or tried and failed to solve it yourself. The biggest gotcha associated with "experts" is that we hope that their experience will magically yield the best solutions. But, by seeking the best from the best, we skip all the pain of learning less than the best and learning about the worst. If we are being told about the best solution, then we are asking someone else to do our work and to think in our place. The worst thing we can do is to accept the best guess from someone else without learning about the problem.

Stop Talking About Solutions

Many of us fall into the trap of talking about our awareness of some solution and guessing at problems to solve, which produces few advantages.

Consider how talking about what someone else is talking about might produce noise. It is often the same with naming a solution, worrying about the future of a solution, pondering if a solution will reach someone else's goal, or debating about how "real" a solution is. Noise, noise, and more noise. Now consider how noticing problems might produce less noise. Problem solving is not noise. Nor is solving a problem at scale. Although that all sounds good in theory, it just so happens that *doing* is a lot harder than *talking*. The simplified and often misguided way we view the role of AI in business doesn't help that dilemma.

American business theorist and professor of organizational behavior at Stanford University Jeffrey Pfeffer discusses how organizations will often talk about mission statements as if they can magically create customers, which is one of the most blatant and common ways to substitute talk for action.[81] Associated with the so-called mission statement problem is the planning problem. Just as some confuse talk with action, many of the same organizations confuse having a plan with implementing a plan. Some may even confuse awareness of a solution with awareness of the right solution. Words are not action, words about action are not action, and not all action is meaningful action.

Many of us believe that group performance or company performance correlates with management's ability to communicate complex technology. We tend to think that asserting solution-focused strategies is as good as having it done already. However, this is not the case. We must notice the right problems to solve. This requires us to first understand if a problem is real, which often involves determining if someone wants said problem to be solved. The right words matter, but they must be aligned correctly with the right action, along with a commitment to that action. This is how we create culture.

---

81. Jeffrey Pfeffer and Robert I. Sutton, *Knowing-Doing Gap: How Smart Companies Turn Knowledge into Action* (Cambridge, MA: Harvard Business School Press, 2000).

Conway's law is an observation that the design of any solution is significantly affected by communication. It is most often associated with software development and organizational team composition but is broadly applicable to any system of communication. The less clear an objective, the more space there is for individuals within a team or organization to fill with their own vision of what problem matters. In the absence of any discernible problem, people may work on better solutions, the wrong problems, or nothing at all. Conway's law highlights that every dysfunction in your team's communication is reflected in alignment, problem solving, decision making, and ultimately your solution.

Build with Humility

The financial incentives to call our work AI is compelling. So compelling that some (but not all) will disguise solutions that are not "real" enough according to some odd external goal. For example, MMC Ventures recently looked at 2,830 European companies that claimed to make use of AI.[82] MMC found that 40 percent of these companies are not actually using artificial intelligence.

However, at this point we should understand that "AI" is not easily defined, so it is not some countable occurrence.[83] In fact, it could be reasonably argued that all and none of these companies are using AI. What the survey completely ignores (which is much harder to estimate) is whether these companies are solving a problem, solving a problem of importance, or solving any problem at all. Consider that one company surveyed said it was gathering "data and knowledge to enable the implementation of AI," whatever that means.

---

82. Unfortunately, this research can no longer be found and can only be stitched together from coverage of the research. All of this is to say that the research was pulled for cause.

83. A Kurenkov, "Please Stop Saying 'An AI,'" https://www.skynettoday.com/editorials/ai-definition/.

What the report is saying is that early-stage companies are lying to investors by claiming "AI" when they're actually using humans. Perhaps the survey really highlights that investors expect companies to lie. Or maybe they don't mind being lied to? What the survey doesn't do is make any attempt to determine whether these "pretenders" are solving a problem. Instead, the survey focuses on solutions measured against someone else's name for their solutions.

There is something disingenuous about secretly relying on humans. Alison Darcy describes secretly relying on humans instead of machines as the "Wizard of Oz design technique."[84] Darcy, a psychologist and founder of Woebot (a mental health support chat bot), explains that the Wizard of Oz design technique is the simulation of the ultimate experience, rather than the design of the actual experience. You may already be aware of the irony of a company that espouses artificial intelligence but does not pursue any facsimile of the technology. Such a company may have a case of Cobbler's Children Syndrome, named for the children of the shoemaker who go without shoes. The fable underscores how noisy words are and paradoxically how noisy even action can be (after all, shoes are being made, just not used).

However, most solutions start off in this way and many businesses hope to scale solutions over some period of time. At this stage in the book, you're probably well aware that solutions have humble beginnings or at least should. Many solutions will rely on the natural intelligence of humans as businesses figure out product-market fit. Keep in mind that premature acceleration is the number one definable cause of a start-up's death and is responsible for the failure of 74 percent of tech start-ups.[85] Premature acceleration is when a business focuses on one dimension of the business and advances it out of sync. The same can be true when we begin to focus on a solution that is out of sync with customer acquisition.

---

84. *See* https://www.theguardian.com/technology/2018/jul/06/artificial-intelligence-ai
-humans-bots-tech-companies.

85. https://www.entrepreneur.com/article/245603.

The assertion by Darcy is part of a much larger techno-optimism built upon normative thinking about artificial intelligence, because the Wizard of Oz design technique ignores that many problems are only solvable with time even though the beginnings are humble. That is because you will need to demonstrate value right out of the box and be competitive against competitors and incumbents. Getting a customer is the first step. But retaining a customer requires demonstrated value, which means you might automate processes using heuristics, simple models, simple rules, and knowledge bases. And you may provide human operators to perform the rest of the work while simultaneously collecting data and building robust data pipelines to train your machine learning solution.

To be sure, your first solution will often be a quick solution, which is unlikely to be your best solution. Google engineer Martin Zinkevich explains in the "Rules of Machine Learning: Best Practices for ML Engineering" that if you think that machine learning will give you a 100-percent boost in performance, then a heuristic will get you 50 percent of the way there.[86] Heuristics, simple models, simple rules, and knowledge bases are ways of putting aspects of a problem back into a solution. Zinkevich's first rule states you should not be afraid to launch your product without machine learning. Despite what Darcy suggests, the Wizard of Oz design technique is a good design strategy. In fact, Zinkevich's third rule of machine learning is to choose machine learning over complex heuristics. In other words, rule number one reminds us to not worry about nonscalable solutions that are unmaintainable if the worse solution gets you out of the door. Once you can acquire a customer and demonstrate value, then you may seek to scale (as rule number three reminds you) to replace nonscalable solutions with scalable ones that are easier to maintain.

Ultimately, good problem solving and good design strategy minimize the need for fully functional prototypes in the early stages of a project. You should have no objection to the Wizard of Oz design technique in the early stages of a company or project when it's adopted in earnest. We should

---

86. Zinkevich, "Rules of Machine Learning."

not be wary of humble beginnings or afraid to discard our first solution, especially nonscalable solutions, for more scalable solutions. How else can you understand why any solution is better than another if you don't have a worse solution to show that you tried to solve a problem in the first place? American computer scientist, and winner of the 1999 Turing Award, Fred Brooks, wrote in *The Mythical Man-Month* that "in most projects, the first system built is barely usable," in part because you are still cutting your teeth on the problem and often learning that your problem was not exactly like you thought.

To be sure, some so-called AI companies do choose to hide behind the amorphous name as a marketing ploy without any ambition or desire to scale. Companies simply protect nonscalable elements by adding more nonscalable units to their solution. However, most all projects will survive their early stages using the Wizard of Oz design technique.[87] The question is: Can we live with the self-delusion that comes from hiding behind the curtain of techno-optimism?

In other words, the Wizard of Oz design technique is a critique that suggests that human workers are problems. This is only true if you are an insider or techno-optimist. Ultimately, the workers behind the curtain, who we are told to pay no attention unless it is to criticize them for being human, should be recognized for their work. In our collective haste to replace superior intelligence with an inferior artificial intelligence, we sometimes forget the humanity of our work.

Critiques such as the Wizard of Oz design technique reflect techno-optimism built upon normative thinking about artificial intelligence that suggests that our solutions can never be real enough unless they render humans obsolete. Criticizing others, as Darcy does, for not using AI that is not AI because there is no AI, while she claims to use AI that is not AI is

---

87. Early-stage ventures will combat issues like the cold-start problem. The cold-start problem in machine learning highlights lacking something in the early stages of problem solving, for example, for supervised machine learning that includes data with labels and known classes and for unsupervised learning that will include knowledge of the right features, number of clusters, or the right hyperparameters.

pretty standard in AI. In other words, disparaging human workers, as the Wizard of Oz design technique does, is not good design, nor is it decent.

Ultimately, the critique encourages the complete replacement, or at least marginalization, of humans at early stages of problem solving. It ignores the value of humans within a solution. It ignores the value of the humans who provide these solutions, and it encourages future automation of the economy. Just keep this in mind. The best way to improve our situation is to improve the situation of all others. That means that our work does not have to be destructive or endorse such reckless technochauvinism to be regarded as real. The real impact of our work is reflected in the problems we solve and the positive impact on all people.[88]

---

88. Well, not "all people." Only something like AGI would have the opportunity to impact all people. Certainly, your people though. That is, your shareholders, employees, customers, and users.

Conclusion

If you meet the Buddha, kill the Buddha.

—Linji

A s W. Edwards Deming is credited with observing, almost every company is designed to deliver the results it gets. If you design your goal to be "AI" then you may miss your larger target of problem solving and customer acquisition and adopt the goals, values, and culture of someone else. American author and former dot-com business executive Seth Godin says it's comforting to use someone else's goals to guide our work as it lets us off the hook.[1] But the only way to do our best work is to realize that it's important to own our priorities. Perhaps you don't have to kill the Buddha on the road, as Linji says, but we should go around him. Otherwise we are looking outside ourselves, making the mistake of trying to follow another on their journey. Going around the Buddha means destroying the notion that any authority outside of ourselves can be our master. If you have the wrong goals, values, perspectives, culture, communication, or attitudes, you will not find or solve the right problems.

---

1. Seth Godin, "Priorities," *Seth's Blog*, August 4, 2019, seths.blog/2020/01/priorities/.

Being AI-first means doing AI last. Doing AI means doing it last or not doing it at all. The reason is rather simple: Solution-focused strategies are more complex than problem-focused strategies; and solution-focused thinking ignores the most important part of business, which is the problems they solve and the customers they create.

Keep in mind that solution-centric thinking results from the following:

- Focusing on what our solutions ought to be rather than what they are.
- Focusing on the impact of future solutions rather than the future impact of today's solutions.
- Conflating our goals with the goals of others.
- Focusing too much on abstract problems with some arbitrary solution or focusing too much on someone else's problem and ignore your problems. The former is solution solving and the latter often means we are working problem solving backward and finding problems to solve in the context of someone else's solution.

Do not define your solution. The search for analytical exactitude in verbal definition will not lead to economic progress. Ignore recycling glib, textbook definitions of artificial intelligence mainly because consumers don't care about textbook definitions. Customers care about themselves. If you want to make your life better, make their lives better. Help them accomplish their goals in a better, faster, safer, or cheaper way. They're generally interested in a value proposition that contains problem-specific information, not in a definition of intelligence. Your journey starts with more comprehension of problems, not the names or definitions of solutions. Besides, creating definitions for our solutions means we are creating external goals for them, which is nonsense.

External goals produce solution arguing and the strange persistence of the question of whether our solutions are real. External goals are explicit when we define artificial intelligence. By defining AI, we add external goals to our solutions that they do not need. Solutions need goals that are aligned only with problem solving. There is no need for a solution to have any goal that a problem or customer did not grant. Ultimately, external goals for our solutions create an environment of advocacy for solutions, a place where we find good problems for our solutions to satisfy the external goals while invariably failing to satisfy the internal goals of problem solving.

Own the name of your solution. Using someone else's name is always tricky because it often comes with their goals. It is important to remember that your goals have to be your goals. Naming a solution AI that is not AI often has a negative impact on prioritization because of its external, and sometimes vague, metaphysical goals. Any name should remove the goals of others and ultimately avoid the trap of following another on their journey.

Stop arguing about solutions. There is no need to argue about whether your solution has met someone else's goals. That's because your solutions do not need goals that a problem did not grant them to have. Ultimately, a solution can only be shown to be fake after it has failed to solve a problem, your problem. A solution is not real based on whether it has a name; is cognitively, psychologically, or neuronally plausible; what it looks like; or if it is or is not intelligent. Furthermore, criticizing others for not using AI that is not AI because there is no AI, while you claim to use AI that is not AI is nonsense. Argue about a solution if you must but always conditioned on a problem and a customer.

Focus on today and ignore tomorrow. Although AI has always been an area of research with two eyes on the future, it doesn't mean business should keep two eyes on the future. Recall that living in the future produces horizon-looking dynamics like the AI Effect, which views solutions as being stuck in the present, and explains why they are evaluated against what someone else hasn't done. This techno-ideological view is a view from nowhere and is dangerous for most businesses.

To be sure, business cannot measure progress against someone else's goals, let alone the future, only against where you started solving a problem or where someone else gave up solving a problem. Ultimately, the present provides us a future opportunity, not the other way around where the future provides us an opportunity today. Keep in mind that futurists know nothing more than the rest of us about the future; they simply possess poorer judgment. Whereas these groups are horizon-looking because they live in a world where they can't get hurt, businesses can be hurt by ignoring today and ruminating exclusively on tomorrow.

Problem solving versus solution solving. Problem solving provides the most useful stopping metric. Problems are first solved when a solution creates a customer. This is product-market fit. It does not mean we stop solving. It means we have created a minimum viable solution. Problem solving is superior to solution solving because solving solutions rarely provides the same type of alignment or stopping metric. This is part of the reason why there are so many variant solutions. The horizon is so far off in the distance that there is no place to stop.

Awareness of a solution is rarely the goal. Awareness of a solution produces confirmation of a solution (a type of solution envy). The cost of being aware of someone else's solution is the cost of ignoring your problems in exchange for someone else's problem. What awareness really underscores is a natural tendency to prioritize solutions over problems to decomplicate problem solving to a single solution or to a single metric because it makes our world seem more manageable and controllable. However, any advantages that come from this approach are short-lived. Invariably, we realize that our world is much more complex than someone told us, and our best work cannot be done by anyone else. By oversimplifying our world, we create second-order effects that create new problems. One such second-order effect: the problems left orphaned while we create the wrong alignments.

Problem solving requires the right attitudes. Solution envy becomes a de facto standard in many organizations: "someone else has built it" rapidly becomes "I need it." However, this standard never supports goal alignment. It's equivalent to "if it has been built, it must be built." In other words, everything already done is true, and what has not yet been done is untrue.

Problem solving requires us to guard against validating claims based on our standards. An invalidated claim should not be regarded as false simply because our standards are too high, and a validated claim should not be regarded as true when our standards are too low.

Solution envy inevitably leads to solution confirmation bias. This bias causes us to find problems that support our solution-focused strategies. For example, if you tell someone to "do AI," they will likely find problems good for some arbitrary solution to satisfy the requirement while ignoring real problems. In other words, the hazard of this sort of biased thinking is we seek evidence to confirm a solution rather than information related to a problem. Problem solving always begins with a problem. Fall in love with a problem, not a solution. If you love your problem, then you can probably trust that it is real. If you love your solution, it probably isn't.

Solution pride is a sin for problem solvers and it's the wrong attitude. Pride of a solution leads to self-praise and self-importance, which subverts problem solving and those impacted. Solution pride causes us to notice the right problems for our solutions and prevents us from noticing the right problems. Pride of a solution often prevents us from throwing a solution away when it has failed to solve a problem. Or it causes us to create more complex solutions when they are unnecessary. Do not love your solution, because it does not love you back.

Small over large. Before you think about the biggest solution, think about the smallest one. There is always time to go big, but starting big will usually exhaust your resources. Do not be scared of solving a problem with a simple solution or problem solving in small steps. Although too simple is often shown to be wrong by insiders based on their values, a real-world solution can only be considered too simple for a problem after it is shown to be wrong. Solving a problem is the goal, not the size or looks of the solution. Remember, only trust those with dirt on their hands. All others are certainly wrong.

Good over perfect. Real-world solutions do not have to have everything. They only need what is required to solve a problem. There are so many problems that did not get solved because someone wanted to do too much but had little, so they did nothing. To make progress on problems, we need to start by accepting imperfection and making do with what is available, unless we have to be perfect. Then you should think about making your problem perfect for your solution.

Lastly, do not let someone else's goals define your business or tell you what problems are most worth solving. Those who are least powerful and least rich are often those whose consciousness is structured for them. Good luck!

Acknowledgments

First of all, I'd like to thank my wife, Nicole, for supporting me and for tolerating how tyrannical I've become about problems and solutions.

Also, a special thanks to Jamie Dos Santos, founder of Cybraics. Without her support, this project would not be possible.

Thanks also goes to Will Grannis, founder and managing director at the office of the CTO at Google; Abe Usher, co-CEO of Black Cape; and Lt. Col. Andrew Wonpat at the JAIC.

Additional thanks to frolleagues who have taught me valuable lessons and helped me along my journey. Thank you, Marvin W., Oscar W., Nathan D., Shrayes R., Jonathan T., Michael C., Kaska A., Travis C., Dmitri A., and Isaac F.

Finally, thank you to everyone on the publishing team at BenBella with special thanks to Matt Holt and Katie Dickman.

Index

A
ablation experiments, 224
academia. *See* insiders
academic problems, 65. *See also* insiders
acceleration, premature, 233
accountability, 195. *See also* responsibility
accuracy, 143. *See also* performance
Acemoglu, Daron, 216
action, meaningful, 231
activities, 209
adversarial data sets, 149, 191
aesthetics, 82–84, 87–88, 93
agency, 124–125, 146, 195
aggregation, wisdom and, 85
AGI (artificial general intelligence). *See* artificial general intelligence (AGI)
agile, 55n
AI Alignment Problem, 22n
AI Effect, 26–27, 95
AI Index, 100
AI Speedometer, 100, 102, 103
AI Theater, 158, 161, 163, 164
AI-first, 1, 2–3
Alexa, 29, 36
AlexNet, 66, 71, 101
algorithms
    comparing, 76–77
    data and, 59, 60, 61
    rule-based, 225
Alibaba, 27–28
alignment, 22n, 25–26, 77. *See also* goal alignment; misalignment
Alphabet, 1, 64. *See also* Google
AlphaGo, 8n, 19–20, 21, 22–23

AlphaGo Zero, 71
Amazon
    Alexa, 29, 36
    Mechanical Turk, 144
    recommendations, 29
Amode, Dario, 71
AmoebaNets, 71
anchoring, 46–47, 50
Anderson, Joel, 27
Anderson, Monica, 26, 80n
ANI (artificial narrow intelligence). *See* artificial narrow intelligence (ANI)
anthropocentrism, 14
anthropomorphism, 14, 51n, 85, 158–159
    of artificial neural networks, 81
    of solutions, 146–156, 158, 161
Apple
    Face ID, 36
    Siri, 28, 29
applied problems. *See* outsiders; problems, real-world
Architects of Intelligence (Ford), 111
artificial general intelligence (AGI), 35, 37–40, 41, 43, 46–47, 112
    attempts to reach, 102
    DARPA and, 112n
    DL and, 119–122
    existence of, 44
    lack of consensus on problem, 111
    need for moonshot, 110
    OpenCog, 160–161
    predictions about, 106
Artificial General Intelligence (Goertzel and Pennachin), 38

About the Author

Photo by Jared Wolfe

Currently, **Rich Heimann** is Chief AI Officer at Cybraics Inc. Cybraics is the first fully managed "AI" cybersecurity company. Founded in 2014, Cybraics operationalized many years of cybersecurity and unsupervised machine learning research conducted at the Defense Advanced Research Projects Agency (DARPA).

As Chief AI Officer, Heimann makes sure Cybraics doesn't *do* AI in the strict, narrow academic sense. That may sound odd, and while Cybraics's solution sits at the intersection of several subfields of AI such as multitask learning and meta-learning, it is not aligned to any external goal of AI; rather, the solution is aligned with how the team has framed the cybersecurity problem. In this sense, AI is a placeholder to make that complex problem of cyber more accessible and easier to manage for humans.

Heimann is former chief data scientist and technical fellow at L-3. Heimann has also been a performer on DARPA's Nexus 7 program, Naval Research Laboratory, Pentagon, and continues to consult with the industry and the Department of Defense. His perspective on problems and solutions is influenced heavily by his background, which is statistics. Statistics is almost entirely applied. In this way, artificial intelligence is very different from statistics. You'll never hear anyone in statistics ask whether or not a statistical solution is "real" or see them create external goals for their solutions.

While experience has taught Heimann that solution-focused strategies are generally wrong, AI-specific strategies are especially wrong for business. This book is his attempt to make sense of AI and specifically what it means to *do* AI for business.